Science, Reason and Religion

edited by
Christopher Macy

Prometheus Books

Originally Published as *Rationalism in the 1970s*, 1973
by Pemberton Books for the Rationalist Press Association.

ISBN 0-87975-027-8 paperback
ISBN 0-87975-028-6 hard cover

Made and printed in Great Britain by
The Garden City Press Limited,
Letchworth, Hertfordshire SG6 1JS

Contents

D. J. STEWART

What is the Message?

> This world can be saved from political chaos and collapse by one thing only, and that is worship. (Temple, 1941)

THAT ASSERTION, made several decades ago by an Anglican archbishop, is an example of a message. You will probably be struck by the fact that he seems to have been worried by exactly the same things as those which are still concerning us today. The problems of the 1970s like those of previous decades, are those of political and moral conflict. What is more, we still fear that we may be on the brink of chaos and collapse.

It is also clear that Temple believed that there was an answer to those problems, and that the answer could be summed up in the one word 'worship'. Are rationalists, too, trying to say something to the rest of the world and, if so, what is the main theme of *their* message? At first sight there seems to be a close correspondence between the two cases. If you were to take the Archbishop's message and change just the one word, substituting 'reason' for 'worship', then you would have a fairly satisfactory statement of the rationalist answer to the problems of this decade.

However, the trouble with any answer that can be

summed up in one word is that it takes a great deal of unpacking. Hence my task is to try to say what we mean by 'reason', how this is connected with rationalism and what is the nature of the message.

Meanings of 'reason' and 'rational'

We had better begin by disposing of the idea that the word 'reason' and its derivatives are all just labels of approval, used by us to indicate what we believe. Far too many people say, 'this is reasonable' when they mean, 'this is what I think myself' and 'is it rational?' when they mean, 'is it true?'. In political and religious matters particularly, the adjective 'rational' is frequently used in a way which does not describe any objective characteristic, but merely asserts which side of a controversy the speaker happens to favour. These are not habits of speech which commend themselves to rationalists.

The word 'rationalism' has always caused a certain amount of confusion. It is obviously associated in our language with 'reason' and implies connections with a host of everyday usages. Let us look at a few of them. There is 'reason', or 'reasoning', concerned with logical argument or deduction from premises to conclusions—the process of showing that something is so by starting with axioms and working downwards from them. Then there is the notion of being 'reasonable' in the sense of being moderate or judicious, not a fanatic and sane; or 'reasonable' can have the slightly different sense of something that seems likely—a reasonable belief to adopt. If you talk of a 'rational' belief, you mean that it is justified by reasoning. You can say 'rational people'—meaning non-fanatic, sane or sensible people. 'Rational' behaviour is the sort of behaviour that you expect from sane people. These are all everyday usages of this cluster of words.

'Rationalism' is also a technical term in philosophy. It refers to the attempt to arrive at completely reliable knowledge entirely by the processes of logic. Then again the notion of 'rational strategy' arises in the theory of games as a strategy which is correct in a certain technical sense which depends on the situation.

My argument will be that our use of the label 'rationalism' is related to all these ideas and that it is possible to build up a consistent structure of thought which incorporates all these notions, and which provides a uniquely hopeful basis for tackling all the problems which face us.

Reason as logical deduction

Let us start with one of the more technical notions. 'Reasoning' can be formal deductive logic: deduction from the premises of an argument to the conclusions, by starting off with those premises and operating on them by some sort of substitution or manipulation rules, so that you go step-by-step down to the point where you arrive at a conclusion. What you have done is to begin to construct a system which has the shape of a pyramid, and I shall often refer to this pyramidal structure that you find in a formal deductive system. At the top are axioms, which are the statements you start with; next come all the possible statements you can arrive at by just one step from the axioms, then those requiring further steps, and so on down the pyramid, which is therefore constructed from chains of reasoning. The base of this structure is wide because here there are a large number of statements. You may be familiar with formal deductive systems of this sort from the Euclidian geometry of your schooldays. You were asked to prove riders by taking certain things as given, going through this process and coming out at a conclusion, which was *quod erat demonstrandum.* There are various other deductive systems having this sort of structure and operating in much the same sort of way. In general, you can build such systems of logical propositions and either regard them as of interest in their own right, as when they are pieces of mathematics, for example, or you can use them for other purposes.

Reason in empirical science

We now move on to one such purpose. You can use the pyramidal structure of formal deduction as the skeleton of a theory in empirical science. The main point about science is that you do not start at the top and deduce downwards; you start at the bottom, with individual events which are the observations that you have made of the world, and you move up the pyramid, going to propositions of greater and greater generality, until you reach the propositions of greatest generality at the top—what used to be called 'laws of nature', although the term has somewhat fallen out of use these days and I prefer to call them 'primary beliefs'.

I do not here want to adopt any particular position within philosophy of science and what I have just given is a very brief summary of a general view about these matters. There are all sorts of variations of this and all sorts of disagreements that we could elaborate and study. One way in which you can look at the relation between empirical systems, of the sort that you would call a scientific theory, and logical systems, of the sort that geometry is, is to say that you can regard the whole thing as a species of game, whereby you have symbols, which you choose as a set, and rules as to how you are going to string these symbols together and how you are going to substitute strings of symbols one for another. You play this game with them, using the symbols like counters, and again you get the pyramidal shape of system, which is uninterpreted. Such an uninterpreted system is called a 'calculus', from the Latin for a stone, because people can do this sort of thing with piles of stones. So one way of describing the structure of a scientific theory is to say that you have your pyramid, which is a calculus developed formally by using rules, and you then interpret it to obtain another structure made up of empirical propositions. These enable you to say things about the world that you have observed. The argument that goes on in the empirical structure is all upward. It uses inductive logic and so is not formally valid. Then, alternatively, you can take the same shape of structure as a *logical* system and, by interpreting each stage of the calculus in these two ways—empirical and logical—you can deduce downwards from axioms in your logical system and so rely on the formal validity of this deduction to justify your beliefs in the other, empirical system.

These rather technical matters are the sort of thing that we talk about when we discuss theory-construction in science. Again, I have to emphasize that there are many different ways of looking at these questions and plenty of people will disagree with the account I have given. However, there are certain things that we want to hold on to: the idea of a pyramidal structure and the idea that you can have at least three different sorts of such structure which have the same characteristics. They each have a small top of just a few propositions of great generality and from these you derive others. If you are doing science, you are justified in your empirical beliefs by the fact that what you say is going to happen does happen. You describe the world by means of the theory and you find that this description is borne out by the observations that you make.

You can be interested in the success of your descriptions, or of your predictions of what is going to happen in the future, or your success in prescribing what people should do in order to bring about a particular result. Also, very importantly, it so happens that the fact that you can put your scientific work into this sort of structure, where you have a logical system as one of the interpretations of the things you are using as symbols, makes it seem to you that you understand what is going on. You feel that you have been given an explanation, because you see the logical characteristics of the logical interpretation of the calculus. People differ a great deal, both in their theorizing about these questions and also in the sorts of system that they find satisfying as explanations. If you want to explore this subject further, you will find that Braithwaite (1953) gives a very full account.

Consistency

When we are dealing with this sort of system, we put a very high value on its being internally consistent. We do not expect to find both a proposition and its contradictory occurring within the same system. We feel that there is something wrong if we do. In the case of logical systems, if this happens there are techniques for dealing with it. In the case of branches of science, we are worried and try to resolve the contradiction by developing our knowledge further. I do not want to suggest that it never happens. You do sometimes find whole pieces of science, which are perfectly well consistent within themselves, but which do not seem to agree with each other or do not easily fall into place in a larger system. Again, there are techniques for dealing with this, but sometimes it is regarded as a case for some sort of experimental test situation, where you would regard one piece as being in some way mistaken and the other not. You *can* get situations arising in science where two systems, which appear to be consistent within themselves, seem to contradict each other and yet they both fit in. An example is in the period of physics during which electro-magnetic radiation appeared to be both waves and particles at the same time. This felt odd, but there were ways of resolving it. Remember this notion of consistency, because it is valuable to us.

Discovery and justification

When we describe any of these activities as operating within a pyramid, starting at the top and working our way down, it sounds a terribly mechanical process, as if you do always just start with your axioms and develop your theorems. This, of course, is not what happens. I am not here describing the process of discovering new theorems or the process of proving the theorems we have been challenged to prove. You will remember that, when you were given a rider in Euclidian geometry and told to prove it, you did not just sit down and mentally 'operate the machine' until you reached the conclusion. There was a great deal of thinking involved, including trial and error, and the recognition of similarities to problems you had solved before. It was not just a mechanical process that you could have given to a calculating machine or to a computer. There are systems where you can have decision procedures, so that you can test the truth of any particular proposition without having to derive a proof, but that is a different process. The context in which you *discover*, as a result of a skilled search, is not the same as the context in which you *justify* the position of the statement in your system, either by formal proof or by a decision procedure.

I can take another example of the same thing. A scientist may engage in all sorts of different activities in an attempt to make progress in his empirical field but, when he writes up his paper for publication, only the essentials are included and he has rearranged them. He shows how you arrange a particular test situation, where you have a hypothesis that can be tested or two contradictory propositions, only one of which can be true, and can set up an experiment so that one or the other is going to be supported, and so forth. Only very rarely do things happen like this in the actual scientific world, but this is again because we need to talk about the distinction between the process by which you *discover* things, which can be a very random matter and involves creativity, and the process by which you *justify* the propositions you are asserting.

The search for certainty

One of the most important phenomena in the progress of human understanding was the search for certainty. It was felt that there must be a way by which you could establish knowledge which was absolutely reliable and that the way to achieve this was, somehow or other, by the operation of reason. One philosopher who took this view was Plato, who thought that the sort of truth you got from mathematical and logical work was somehow superior to the sort that you got by observation. He was one of the first people who tried to establish a philosophy in which certainty came from reason and was regarded as preferable to the sort of knowledge that comes from experience.

There have been many attempts along these lines. Those philosophers who have built up systems constructed by formal deduction, have always felt that somehow or other they have to choose the axioms at the top of their systems by some means so that these axioms are, without any doubt at all, true. The problem is, how do you obtain such reliable axioms? Somehow, they have to be self-evident. You may remember that Descartes was worried about this problem. He felt that the experience you get by seeing things around you could be misleading. You could be mistaken about it. It is not absolutely certain, he argued, and the sort of knowledge that you get from self-evident axioms or from mathematics is somehow different in that it is more reliable. I think that Plato, Descartes and others who took this line made a mistake because, just as one could be dreaming that one is sitting by a nice warm fire (Descartes, 1641) so one can dream that one has carried out a calculation and got a particular answer. Have you ever got to the end of a calculation in arithmetic and felt that you were not absolutely certain that this was the correct answer, not to the same extent that you were certain you were sitting at your desk? Are you more certain that the square on the hypotenuse of a right-angled triangle is equal to the sum of the squares on the other two sides, than that you are really reading this book and not just dreaming? There is something wrong with that traditional argument and we are not now disposed to accept it, even though the philosophers who adhered to it were also called 'rationalists'.

This general position in philosophy was called 'rationalist' because it tried to rely on the use of reason entirely—using it to deduce from self-evident axioms. We are the inheritors of this tradition, but have abandoned this aspect of it.

The line that we would now take is that philosophy began to make progress when the search for certainty was abandoned—when it was seen that you could accept or reject the *whole* of a deductive system according to what happened at the bottom end of it. You could take a scientific theory and accept it if the observations fitted or modify it if they did not, and you did not have to worry about having totally reliable axioms at the top end of the system. You could justify the axioms because you could justify the complete theory—which in turn you could justify because the bottom end fitted with experience.

Abandonment of the search

The abandonment of the search for certainty, and the substitution of an attitude whereby you neither ask for absolute certainty nor expect absolute truth, but are prepared to accept the best that is going for the time being and to give it up later if you find something better—always ready in the meantime to look for something better in an active way and not to be worried by the need to change your mind occasionally—was an extremely important change in the way that people went about attaining knowledge. The fundamental conception of what it means to have knowledge was altered by that change of attitude.

A professional scientist, in theory at least, operates within rules that are an extreme version of this approach, whereby there are for him *no* propositions which must always remain true. There are no propositions at all which he is not under certain circumstances prepared to give up. What is more, he will be able to state for you, in theory, the logical status of these propositions, and will be able to tell you under what circumstances he *would* be prepared to give them up. This is related to the development in scientific thinking whereby you can have empirical statements which have a very high degree of probability or a very low degree of probability, but which cannot be either completely certain or completely ruled out. However much evidence you get in favour of an empirical proposition, the probability can continue to increase but can never actually reach unity.

We are building up a structure in which there arises the notion of *justifying* one's beliefs, and the sort of justification depends on the sort of belief you are dealing with. If you are dealing with a logical proposition, then your belief is that it is a theorem in a deductive system. Your way of justifying this is to carry out the calculation and demonstrate that it is indeed a theorem. On the other hand, if it is an empirical proposition, your way of justifying it is going to be in terms of *obtaining evidence* for it. There are some difficulties about this, because your scientific theory remains with you as long as the observations that you make are in conformity with it, and your confidence in it can increase the more such events occur to you; but no number of positive instances can ever act as a proof of this theory for you. There is always the possibility that you will have to give it up. There is a lack of symmetry about this situation, because a contrary incident, an observation which does not fit the theory, is much more serious to you—you know that something is wrong. One contrary incident immediately sends you back, looking at the theory to see what is wrong with it.

I think it is worth mentioning that, if you have derived a prediction within a scientific theory and events do not work out the way you expected, you cannot be sure that you merely have to give up the little bit of the theory nearest to that event. You cannot say where in your theory you have to make a modification. Strictly speaking, what has happened is that the whole of your scientific understanding of the world has crumbled about you and you have to start again. Obviously you will use nearly all of the same propositions again and will build a new theory very like the old one, but the necessary change could be anywhere. What is more, there is a large degree of choice available to you. If there is a proposition in your theory that you rather like, you can hold on to it and make the change somewhere else. Events do not force you to make the change in one particular place, so you can hold some of your beliefs incorrigibly and modify your theories in quite complicated ways in order to save the beliefs you want to save. The scientist has more choice in the sort of theory that he is producing than perhaps some people think.

Are there no eternal truths?

When we say that there are no eternal truths at all, do

we include the assumptions of the scientific method itself? Are there perhaps some assumptions, after all, that are constant points in the development of our understanding of the world? What about the rules by which we build up our theories? Are these constant? I suggest to you, firstly that even these are not constant. Scientific method itself evolves. Even the parts that you would think very stable —logic and the foundations of mathematics—have evolved enormously over the last few years. Our understanding of the philosophy of science and our more sophisticated approaches to the scientific method are certainly developing, just as is any other branch of knowledge, so this is not a fixed point. But perhaps you could say that, just as any other branch of knowledge is evolving with a recognizable continuity, so also is scientific method. That is one sort of answer.

The second sort of answer we can give to the question as to whether this is a sort of eternal truth is that there is no particular reason we can give—no justification—for choosing one set of rules for our theory-building rather than some other set. We do have our favourite set of rules and all scientists use more or less the same set; but, if somebody comes along and insists on using different rules, we might think him odd but we could not produce any reason why he should not do it. It is a situation rather like that which happens when you are playing a game of Association Football and somebody picks up the ball and runs with it. He is breaking a rule and you can penalize him—but only so long as he wants to remain a member of your club. You can only say, 'We are not going to let you do that here. If you do not intend to keep to our rules, then you have invented a new game. Why not call it 'Rugger' and go and found a new football club down the road?' There is a choice of which game you intend to play.

The choice of a set of rules is simply a matter of making a decision to accept them. Within the system of these rules you cannot have a justification of your decision. This is related to the philosophers' problem of the justification of the inductive method, about which there is still a great deal of argument. Personally, I prefer the way of justifying induction which says in effect that induction is the only method we have, of which we can say how we are going to use it in the future. If it does not work we are lost, because we do not have another method. Therefore the rational strategy is to use it.

There can be other rules; there can be other logics. You occasionally meet people who appear to be using a different logic from the standard one of the mathematicians and the scientists. For example, if I were to say here that everything said here is true, you could look at this and believe that it must be true, because it says that it is. Such a process of reasoning would be invalid according to the usual rules of logic, but I do not know of a reason why there should not be founded a club for operating according to these different rules. As a matter of fact I think that such clubs do exist.

I have talked about the rational justification of beliefs and have discussed the way in which one rationally justifies logical beliefs and empirical beliefs. Now we must leave belief and move on to behaviour.

Rational justification of behaviour

Not only can one rationally justify one's beliefs, as I have described in the case of logic and science, but the same sort of process is also applicable to behaviour. Reason can help to tell you what to do, as well as what to believe. Here again you have a system with a pyramidal shape. Just as I can say to you, 'Make sure that your beliefs fit into such a system,' so I can say, 'Make sure that your behaviour does the same.'

You cannot logically deduce what to do in a situation from a pure description of that situation. From a set of facts only, you cannot rationally justify an action. That is a piece of logic which was first suggested by Hume. But, if you have available to you a mixture of facts and purposes, then you *can* deduce what to do in that situation. From a mixture of purposes and facts you can build a structure whereby what you do is rationally justified because, in the situation described, it leads to some purpose being achieved, which in that situation leads to another purpose being achieved, and so forth. You can do a purpose analysis which typically yields a pyramid structure having superordinate purposes, from which you can deduce other purposes. You can justify doing things at the bottom because they lead to something higher up, and so on right up the structure. I have used 'purpose' in preference to 'goal', but I do not want to suggest that there is no way of subjecting goal-directed behaviour to causal analysis. On the contrary, this is the chief topic of cybernetics, which is my main area of work.

There is a, partly technical, use of the term 'rational behaviour' for behaviour which, in a given situation, enables you to achieve your purposes. This definition fits your everyday, intuitive notion of what is rational behaviour. Irrational people are those who do things which, in the circumstances, do not lead to the achievement of the purposes they have.

Rational versus pragmatic justification

A particularly interesting form of purpose analysis is that of a competitive or conflict situation, where you have some people trying to achieve a purpose, and other people trying to achieve a different and incompatible purpose. When you are in such a competitive situation, how can you ensure that the outcome is nearest to what *you* are trying to achieve? The mathematics of this is called 'theory of games', and it is in this that 'rational strategy' is a technical term. A useful account of the subject has been given by Williams (1966).

May I make a point here about what is meant by the rational justification of a particular act? Supposing that you are a gambler who accepts a bet at unfavourable odds and it so happens that you are lucky and win, you can take either of two attitudes to this event. You could say that you ought not to have accepted a bet at unfavourable odds. If you keep on doing this you will be ruined. Your understanding of probability theory, game theory and so forth, tells you what you ought to do to win money and this behaviour is not a part of what it tells you. You were not rationally justified in accepting that bet at unfavourable odds. On the other hand, you could take the view, 'Well, it paid off, didn't it? I've got the money. What more justification do I need?'

I suggest that the first is a rationalist attitude and that the second is not. There are many analogous situations, where there is a distinction between a rationalist attitude and what we might call a 'pragmatist' attitude. There is a clear-cut line and rationalists are unequivocally on one side of it.

Can we rationally justify primary purposes?

I have talked about rationally justifying purposes by logically deriving them from superordinate purposes. If we have some grand, overall purposes we can, by the use of rational analysis, deduce what to do in terms of smaller purposes, and I have at least indicated how one can justify pieces of behaviour within a pyramidal system. There is a question that people sometimes ask: 'How can you rationally justify the top end of your pyramid?' I have said a certain amount about the rational justification of axioms in logical systems and the primary beliefs in scientific theories. How about the top ends of practical or purposive systems? The point is a controversial one, but I think it is a fair statement of the logic of the situation to say that there is no way of rationally justifying a choice of particular primary purposes. I am using the term 'primary purpose' to mean those right at the top, which cannot be derived from anything higher up.

Incidentally, when you carry out a purpose analysis, you may find that your purposes are not arranged in a neat hierarchy at all. You may find that purpose *A* can be derived from *B, B* can be derived from *C* and *C* can be derived from *A,* so that you get a sort of circle at the top. Things like this can be very confusing but, generally speaking, you have some things which are equivalent to primary purposes in such a system, and there is no way of rationally justifying the choice of these rather than any others. What does a rationalist do then?

There are one or two tricks you can carry out at the top end of your purpose system. For instance, you can say that it is a sort of primary purpose to want to be able to achieve your purposes; so the situation in which you find yourself may affect your choice of primary purpose if you know that some are achievable in that situation and that some are not. Then you will probably choose the achievable ones in preference to the non-achievable ones.

To avoid having any purposes at all, on the grounds that you could not justify your choice of any particular primary ones, would not be a rationally justifiable position. I suggest that the only course open to you is simply to adopt primary purposes, without worrying whether they can be rationally justified or not. If there is no way in which some can be justified more than others, then they are all equal in this respect and you can choose whichever you like.

It is probably more accurate to say that you will not in practice need to make such a choice, because there will be plenty of factors in the situation which will cause you to have some purposes already. These will be well established long before you are in a position to think about such matters. The most you can hope to do is to tidy up and evolve your purpose structure, in accordance with the rules of rational justification but on the basis you have been given. People come to have purposes as a result of the combination of heredity, physical environment and social environment. They have a particular genetic endowment and a particular sort of upbringing and have had particular conversations with other people, who have been persuading and indoctrinating them. Nobody in practice simply sits down and chooses purposes in a vacuum. Incidentally, Eysenck (1967) has raised some very important questions about the place of indoctrination in a world of rationalists, which so far as I am aware have not yet been answered.

In the tidying-up process of maturity, people attempt to evolve primary purposes in order to justify the subordinate purposes they already have, in much the same way as the primary beliefs of scientific theories are evolved to justify belief in specific empirical propositions. This is not the only close parallel I wish to suggest exists between belief systems and purpose systems. The logical justification I have proposed above for making a choice of primary purposes at all is similar to the justification of induction adopted on page 20.

Reason in ethics

Although primary purposes cannot be rationally justified, philosophers have for a long time been trying to find a justification for them. This will not surprise you; philosophers are, on the whole, rationalists in our sense of the word, in that they are concerned to put forward whole systems which are validly deduced from a few principles, after the style of one of our pyramids. Professional philosophers are nearly all rationalist in this sense and there is no philosophical position more rationalist than another, because they are all obeying these same professional rules. All philosophers who are concerned about the use of reason to justify a particular practical system are worthy of consideration. The area in which they do this is moral philosophy or ethics, and there have been plenty of suggestions made about how you can justify a set of primary purposes by deriving them as equivalent to moral directives, imperatives, maxims, or principles.

Here are a few examples of the ways in which philosophers try to establish systems of ethics. They can try various logical tricks in the hope that they can find self-evident axioms—to find primary purposes which have some technical characteristic within logic which particularly recommends them. Kant is an example of somebody who worked along those lines. Then there are those who say that the world is the sort of place where, if you look around you with a special sort of moral eyesight, you can simply *see* that situations and things possess characteristics which have moral relevance. All you need do is open your moral eyes and look, and you will see what to do. Then there are people who look at the situation and fasten on to a particular characteristic. Spencer, for example, fastened on to Darwinian organic evolution and suggested that there are senses in which we can say that things are getting better in the organic world. We can say that it is getting better in a moral sense as well, and extrapolate this to give us an ethical direction. A recent example of the same type of theory is that put forward by Osborn (1970), who uses a similar argument at the individual level to say that, as people grow up, we can see psychological changes which are a growth towards maturity; as we like the look of this development in some respects, we can pin an ethical value to it as well. One might raise philosophical objections to this along Hume's lines: that such theories try to derive *ought* from *is*.

Another philosophical position about morals, which is often quoted nowadays as a means of justifying particular primary purposes, is utilitarianism. I am far from convinced that this is any better than the others. In fact, it has had a fearful drubbing from professional philosophers and so far as I am aware there is no version of it which is internally consistent, which is a serious objection to it.

All these different approaches are subject to some objections within the world of the professional philosopher. For philosophers, and hence ultimately for us, none of these systems is entirely adequate. We ought to take this fact seriously in deciding how much weight to put on such systems when we seek justification for our own choices of primary purposes.

Is there a rationalist ethics?

A rationalist who adopts any of these philosophical positions does not offend against the principles of rationalism to any greater extent than he might be offending against those of professional philosophy, if the positions happen to be inadequate from a philosophical point of view. The criteria which determine whether or not a rationalist may properly hold a theory are no different from those which are most commonly applied in a university or other scholarly or educational community. If any difference appears to exist, it may be that rationalists strive to apply these criteria more thoroughly. In my view, this is true not only of moral theories but also of political ones and those in every other field of study. In the case of ethics, this leaves us with considerable scope for conflict, because I think it is fair to say that there is no account of moral questions which receives the unanimous approval of professional philosophers.

It is helpful to compare what happens in the case of scientific belief systems. If a scientist comes to believe a new theory, he publishes it and there is implicit in the publication a demand that all other scientists modify their beliefs to conform to it. Being a scientist implies that you have a right to make this demand of your colleagues and that they in their turn have a right to resist. It is essential to the evolution of new knowledge both that all scientists recognize this right of demand in their colleagues and that each one exercises it on his own behalf. The ensuing conflict is a productive collaboration because all the participants are committed to a common set of principles governing the resolution of such conflicts.

Two features of this traditional scientific collaboration are of particular interest to us. Firstly, weight of opinion is not relevant to the determination of truth. As John Stuart Mill (1859) put it: 'If all mankind minus one were of one opinion, and only one person were of the contrary opinion, mankind would be no more justified in silencing that one person, than he, if he had the power, would be justified in silencing mankind.' This maxim is in no sense dependent on feelings of shared humanity. It would be just as valid if the 'one person' were a computer. Secondly, it is not enough merely to avoid silencing a contrary opinion. One also has to listen to it and recognize its implicit demand that one alter one's own opinion. It is not a rationalist attitude to say that anybody can say what he likes provided one does not have to take notice of what he says. People who do take this attitude should note Mill's further comment: 'If the opinion is right, they are deprived of the opportunity of exchanging error for truth: if wrong, they lose, what is almost as great a benefit, the clearer perception and livelier impression of truth, produced by its collision with error.' I cherish that phrase 'collision with error'; it neatly expresses the continual skirmishing which should take place in a rationalist's thinking. For him there is no deep slumber of a decided opinion.

These points are applicable not only to the development of scientific beliefs and theories but also to that of philosophical ones, including those about morals and politics. The rationalist position on these matters is, I would maintain, that each person is entitled to contribute to the eternal collision of opinion, in precisely the same way as he would be in a scientific matter, with the same obligation to make demands and to grant those of others.

This view may be criticized on the ground that questions of morals and politics are more like logical problems than like empirical ones, but my reply would be that, if this is the case, it is also true of parts of science. The rules of conflict in mathematics may differ from those in empirical studies but, even if they do, both sorts are included among those which scientists recognize as governing their work.

Reason in the resolution of conflict

Human beings have many purposes in common and so groups of them can collaborate in the achievement of mutual purposes. If it requires two people to build a shelter, and there are two people who both want a roof over their heads, they can collaborate to build the shelter and then both use it. It does not follow, however, that having shared purposes means that you only collaborate and that there is never any conflict. You can have situations where there is only one small roof and fifty people want to live under it; having shared purposes can be the cause of conflict.

Can we, as rationalists, help with the resolution of conflict that occurs in this way? Yes, we can analyse the situation, using the tools that I have already mentioned —mathematics, theory of games and so forth—and devise a set of rules governing the conflict, in such a fashion that people can either accept the rules or not. In many situations of this sort, it is possible to set up the rules so that people want to stay within the rule structure—do not want to leave the field—and you can often resolve the conflict by doing so.

The theory of games ought to be of particular interest to utilitarians. One of the objections to utilitarianism is that it is very difficult to find a way of measuring the happiness you are trying to maximize. Braithwaite (1955) has analysed what happens in a situation where people are in conflict but each person is able to list possible outcomes in a preference order for himself. (The orders of the different people do not have to be related to each other.) The theory of games can be used to build up a rule structure within which the competitors play to achieve a maximum-happiness solution.

I suggest that it is theoretically possible, in any conflict situation, to construct by rational analysis a set of rules within which people can operate and sometimes resolve the conflict, but nothing that I have said implies that you can always prevent people getting into violent or war situations. You can if you like take it as one of your rules that you want to avoid these violent outcomes. In other words, you can try to set up the rules so that at no time is it to someone's advantage to declare war or to hit his competitor. There is a great deal of such theory available nowadays and conflict resolution has become quite a developed science. If there are conflict situations, then rational analysis can help tremendously. This is an important part of our message. The effect is to replace rhetoric by logic. Just as a considerable advance was made when mankind began to use persuasion instead of force in the resolution of conflict, so now we are proposing that a further advance is long overdue. The time has come to replace persuasion by calculation. Many situations, where it looks at first as if the only answer is a fight, can be made peaceable by setting up rules which the competitors are pleased to play within.

Do not let this frequent reference to rules lead you to think that there is an authoritarian flavour about what I am saying. For the purposes of most of my argument, it is necessary that the commitment to a framework of rules is voluntarily undertaken. I am not talking about *forcing* people to play according to certain rules; that would be an entirely different sort of approach. On the other hand, do not be put off by my talking about 'games'. This is a semi-technical term, meaning any behaviour where competition is governed by a set of rules, and it should not be taken to imply that the activity is necessarily frivolous or trivial. Some of the situations to which the theory of games may be applied are very serious indeed.

Rationalists in conflict situations

What am I suggesting that rationalists do in conflict situations? I am implying that the appropriate rôle is to analyse, by the methods I have discussed, the logical status of the conflict. Firstly, you can get into conflict which is due to difference of belief over calculation. Generally, that sort of conflict does not last long, because you check your calculations and agree them. Alternatively, people can become involved in conflict arising out of a disagreement over an empirical matter. This is the sort of situation you find within science the whole time. The appropriate way of resolving it is to check the evidence or set up an experiment. The rôle of a rationalist here is to arrange the situation so that people do search for the evidence. If it really is this sort of conflict, it does not usually last long once you have the evidence.

Sometimes, conflict which *appears* to be on logical or empirical matters can continue to exist, however much you apply such analysis to the situation. Crawshay-Williams (1957) has made some valuable contributions on what he calls the 'double criterion of empirical judgment'. This work is concerned with the way in which you can resolve certain sorts of conflict, when they do not appear to be any of the sorts that I have so far mentioned.

As you will have realized, there are plenty of conflicts which are not of any of these sorts. Quite often you come across a conflict which is neither over calculations nor over empirical matters, but is due to conflict of interests. Fifty people fighting for one roof might be an example of this. There are situations where people's interests are incompatible and they cannot all win.

When this happens, one thing you could do is to explain the situation to them. This may help, or it may make matters worse. It is useful to consider what your own position is. For example, if in a political matter it seems to you that one side of the conflict is right and the

• •

other side is wrong, it follows that, if you fail to analyse it out as a logical or an empirical matter in which the other side is mistaken, then you must yourself be a member of one interest group. It could be to the advantage of your group to make public the results of your analysis, or it may be to their advantage if you pour as much confusion over the situation as possible. I certainly do not want anything I have said to suggest that rational analysis always helps you in your own position if it is made public. It could make your position worse, and I cannot see any rational justification for making your own side weaker by illuminating the situation, when you could do better by confusing things. This incidentally leads to the useful principle that the side which is creating most confusion is probably the side which is in the wrong.

If you see such a fight going on and you are not in one of the interest groups involved, there is no rational justification for joining in that fight, unless you just happen to like fighting. If neither side is right, there is no justification for joining in on one side. The rationalist rôle in this situation must be to act as a sort of referee, and I have suggested how this referee rôle might be filled, in what I have technically described as a game-theoretic fashion.

Rationalists as counsellors

Inevitably, when we are taking part in the discussion of any sort of policy—whether it be party-political, in business, counselling an individual who has personal problems, or anything else—there is always a mixture of our own purposes with the advice that we give. If we are professional givers of advice, we may regard it as part of the professional rules for us to withhold our own purposes and concentrate on giving advice which consists purely of the cognitive analysis of the client's purposes, showing how he may best achieve them. But there is always a directive element, even here. If a client comes to you and says, 'I want to blow up your house. Can you give me technical advice on how to do it?', you might very well find that your own purposes intrude on your ability to give this technical advice. I suggest that this is always true; teachers, parents, psychologists and everybody else always meet this problem to some extent. We cannot just say that our role as counsellor is to be cognitively analytic and not at all directive. This fact raises problems and there is a great deal of rationalist discussion about the rôle of direction in counselling. The current concern with moral education, and whether it involves moral direction or only cognitive analysis, is a closely related issue.

Rationalism in personal life

So far, I have been talking about the theory of rationalism. The discussion of any set of principles always seems far removed from everyday life, and you will be asking whether commitment to such principles makes any difference in one's approach to personal problems and social contacts. So here are some possible ways in which you might allow rationalist principles to guide your everyday behaviour.

In matters of belief, try to see the logical status of each belief that you hold and to ensure that you have an appropriate rational justification for each one.

Remember that a logical proposition is justified in a different way from an empirical proposition. For every belief, ask yourself under what circumstances you would be prepared to give it up. According to the sort of answer that you give yourself to this question, you will know what sort of belief it is. Many beliefs, that look as if they are empirical, are held incorrigibly, in which case they are not ordinary empirical propositions and you need to work out what their status is. If you find yourself disagreeing with somebody else over a belief, work out what sort of test situation can be set up, if one is possible, or see whether it is one of those special situations in which there is no possible test.

If you ever hear somebody say that he is *absolutely* certain about something, or that he believes some action to be *absolutely* wrong, this can be a warning signal to you. If you get the chance, ask whether there are any circumstances at all under which he would be prepared to change his mind or allow an exception. This will help you to find out the status of his belief.

Recognize that you and other people will usually be inclined to look for evidence supporting a belief, and to overlook any contradicting it. Everybody suffers from selective perception. But remember that refuting instances are more valuable to the development of understanding than confirming instances. You make fastest progress when your expectation is *not* fulfilled and you find that you have been wrong about something.

In action, try to be clear what your purposes are, particularly those high up in your pyramidal purpose structure. Do not allow low-level purposes to become so interesting that you forget to ask to what ends they are the means. Do not fall into what psychologists call 'autonomy of motivation'. Keep asking yourself what you are really trying to achieve. When you are clear what your high-level purposes are, try to arrange your behaviour to be a rational strategy for achieving them.

I cannot give any advice about which primary purposes to adopt, but you will not need it anyway. It is, however, valuable to clarify what they are, if you can. Do not be too alarmed if you find that this is more difficult than you expected. People come to have purposes as a result of a variety of causal influences, and these do not necessarily produce a neat pyramidal system. The task is a little like doing Natural History, where you take a large number of observations and then fit them into a logical structure.

If you find that there are inconsistencies either in your beliefs or in your purposes you ought to be worried. Consistency is not in itself a virtue. It is not the case that, if you are consistent, then you are bound to be right. You could possibly be completely wrong. But, if you are *not* consistent, then you must be wrong somewhere.

Another thing to avoid is the confusion of means with ends. It is easy to turn your purpose pyramid upside down. Do you ever find yourself deciding that you would like to buy something, and *then* thinking up reasons why you need it? Do you ever find yourself wanting to carry out a particular operation and trying to argue what purposes it would fulfil, rather than first deciding the goals and then the means to achieve them? You will find plenty of examples of this 'purpose inversion' in business and in politics.

When arguing in favour of a particular operation, try to bear in mind what you are aiming to achieve, at the highest level you can. Make sure that you have not turned the argument upside down and chosen a general principle merely to justify a specific consequence. Circumstances change and you may find implications that you had not expected. Make sure that you have carefully worded your statement of high-level purpose. For example, in the arguments about family planning over the last few years, the general maxim has occasionally been proposed that a woman has a right to decide how many children she is going to have. This was used as an axiom of an argument for family planning; but, if you later become concerned about the population explosion, you might feel that you did not after all intend to say that a woman had a right to have as *large* a number as she wants to, and your incautious wording of the original principle has recoiled on you.

Who are rationalists?

What sort of people might you expect to be rationalists? What I have been saying implies that they are like the ideal of professional scientists in being very moderate, judicious, open-minded people, who are always prepared to give up their judgments once made, if they see the evidence is against them, and so forth. Probably this is in fact untrue of the actual people involved, although it is a fair description of the rules under which they operate. We have to make this distinction: that you can aim for something without necessarily being yourself a good example of it. You can approve of something without being able to achieve it. The degree to which even a scientist can behave in the way I have described in his own special sphere depends on the development of his skill in that sphere. He has the advantage of training in being rational within this special area. This skill does not necessarily transfer to other areas. You can think of examples of scientists who, when they are outside their special field, seem to abandon the ways of thought and behaviour that are appropriate to them as professionals. Few people get sufficient education in the sort of generally transferable scientific and logical methods of thinking that I have been describing. The sort of skill that a rationalist tries to achieve in himself, and perhaps tries to instruct others in, is a general one applicable to any field of belief or behaviour, and is not necessarily quite the same as that which scientists are required to learn.

I do not know whether it has been established that scientists have a particular sort of personality, characterized by judiciousness and similar qualities. From nothing that I have said does it inevitably follow that this would be the case. Similarly, it does not follow that rationalists, if we gave them psychological tests, would be found to have special qualities of intellect or personality. If people are specially interested in rationality and the teaching of rational modes of thought and behaviour, it does not necessarily follow that they are themselves more rational than other people to start with—although we might hope that contact with this subject-matter would eventually have its educational effect.

Incidentally, all that I have said about scientists also applies to any other people who are committed to belief and to behaviour in accordance with a system of the sort I have described. Scientists are particularly instructive as an example for us, because they are professionally concerned with conflict involving logically justified beliefs, empirically justified beliefs and strategic behaviour, in a fashion which is all obviously governed by their professional rules. But with some modifications the same might be said of lawyers, business men, bookmakers or anybody else whose trade or profession has to be conducted in accordance with principles of rational justification.

It would be quite wrong to infer from all this that there is no place for emotion in the lives of rationalists. Rationalists can be as emotional as anybody else and, with one proviso, there is nothing in their principles which says that they should not be. The one, very important, exception is that emotion has no place in the justification of beliefs or of behaviour. What we usually call 'wishful thinking' in all its forms is indeed a breach of the rules.

Problems of the 1970s

There have been problems in every period of history. Most of them are still recognizable in the things that worry us today. True, there have been changes—some problems have been solved—but it has usually been the case that each solution has generated a new problem. There has been one recognizable trend. Provided that you have no power, it does not matter too much if you do not know what you ought to do; but the more and bigger things you are able to do, the more important it is not to make a mistake. The explosive increase in scientific and technical knowledge, bringing with it understanding of *how* to do things more efficiently than ever before, has shown up only too clearly how far behind we still are in our understanding of *what* to do.

In other words, the one problem which underlies all the others which face us today is an imbalance between on the one hand our progress in mathematics, logic, physical science and technology, and on the other our failure to make comparable progress in applying the same sort of thinking to purposes, collaboration, conflict and ethics. Particularly significant is the failure of even such knowledge as has been won in these fields to find its way into the education, either of the general public, or of those who have special concern with political or moral matters.

The other major change in the nature of the problems facing us over the last few decades has been the vast increase in the size of all the parameters involved. One immediately thinks of the exponential rise in population which has been going on for centuries, but many other factors have also increased in similar fashion. Not only do more people exist now, but you can also talk to more of them simultaneously as a result of the improved speed and range of communications, and you can meet more of them face-to-face by being able to travel faster and further. Hence the effect of any political decision you make is greater, and so are the complexity, speed and size of any reaction to it. The use of energy and with it the ability to exert destructive power have vastly increased, so the penalty for making a mistake can now be enormously greater than ever before.

If you look at graphs showing any of these important parameters of human life, you will see that they all have a characteristic exponential shape (Benn, 1971; Parsons, 1971). A special feature of the problems of the 1970s is that they have all been made more dangerous and more urgent by this exponential increase.

It follows that, in so far as the answer to these problems is likely to lie in any one all-embracing strategy, it must propose a means to redress the imbalance between people's knowledge of *how to do* things and their knowledge of *what to do*. Fortunately, we have a model to follow. In the history of mankind's understanding of the world, the explosive increase of knowledge was associated with the development of scientific method—with its special technique for bringing together logic and experience, with its respect for evidence and, perhaps most important of all, the particular kind of cognitive humility which characterizes the abandonment of a search for eternal truths. When mankind gave up the quest for absolutes, became satisfied with something more modest and was prepared to build a body of knowledge from beliefs that were always tentative—that might always be given up in exchange for better ones—then scientific understanding blossomed. So why not let us now use this as a programme for achieving a similar blossoming in moral and political matters?

Some of the developments that I have been describing form the foundations of such a programme, and we can contribute to the solving of the problems of the 1970s by helping to spread knowledge of them. In the past, rationalists have been particularly concerned to bring knowledge of new scientific discoveries to a wider public; for example, they can be proud of having once earned the nickname, 'Darwin's Witnesses'. In the 1970s, the new discoveries which will most need to be expounded will be those concerned with the resolution of moral and political conflict by rational means.

REFERENCES

BENN, A. W. (1971), 'Machines and people', *Towards an open society*, London: Pemberton

BRAITHWAITE, R. B. (1953), *Scientific explanation*, Cambridge: Cambridge University Press

—— (1955), *Theory of games as a tool for the moral philosopher*, Cambridge: Cambridge University Press

CRAWSHAY-WILLIAMS, R. (1957), *Methods and criteria of reasoning: an inquiry into the structure of controversy*, International Library of Psychology, London: Routledge

DESCARTES, R. (1641), *Meditations on the first philosophy*

EYSENCK, H. J. (1967), 'The place of indoctrination in the world of Rationalists', *Rationalist Annual*, London: Pemberton

MILL, J. S. (1859), *On liberty*

OSBORN, R. (1970), *Humanism and moral theory*, London: Pemberton

PARSONS, J. (1971), *Population versus liberty*, London: Pemberton

TEMPLE, W. (1941), *The hope of a new world*, London: Macmillan

WILLIAMS, J. D. (1966), *The compleat strategyst*, New York: McGraw-Hill

CHRISTOPHER EVANS

Rationalization, Superstition and Science

In writing about superstitions in the twentieth century, I intend to give something of the background to my interest in the subject and something of the background which I believe is responsible for the present growing interest in peculiar cults of one kind or another. Perhaps it is strange for a psychologist and a scientist to be interested in such things, but I have been struck by what I have learnt about modern cults and superstitious beliefs and even about the evolutionary stage of religious beliefs in forms like Spiritualism.

Coming to this subject as a sceptic, I have been astonished by the tremendous scientific interest Spiritualism has aroused. By scientific interest I do not mean only that the odd scientist has said, 'let's consider the peculiar alleged phenomena of Spiritualism or psychic research or parapsychology' or whatever he might have called it. A very considerable number of scientists, many of them men of world prominence, have become deeply involved in Spiritualism, in psychic research and in parapsychology. It was thinking about this which set me on the track which led to writing my book, *Cults of*

Unreason,[1] and another one that I am presently working on. The latter, titled *The Death of Ghosts,* is about the growing interest in ghosts during the nineteenth century—and by ghosts I mean the objective appearances of dead people—and the development of this interest up to the present day.

Everyone wants to know what the world is about: we want to know what we are doing here; what is going to happen to us; what the purpose of life is, and so on. Those of us who have thought about these problems —philosophers, rationalists, scientists—realize that questions of that kind are not always meaningful when posed in that form. But for most people, some kind of answer to simple questions of the 'why are we here?' variety is required. Until some time during the nineteenth century the best answers to these questions were given within the framework of religion, whether orthodox or unorthodox. If we go far enough back we find that religious beliefs—I am using this term in a very broad sense—helped to answer such questions as 'what is the sun and why does day follow night?'. With the evolution of natural science and the evolution of scientific observation, people gradually found that the answers offered by orthodox religion were progressively less valuable. In a sense, something was taking the place of religion, and that was the tradition of scientific belief. Suddenly astronomy was able to answer the questions, or some of the questions, about what the universe is, how far the sun is from earth, why it gives out heat and so on. Even the limited advances in physiology, psychology and anthropology helped to tell people something about the nature of man.

[1] *Cults of Unreason* by Christopher Evans (Harrap £3).

Science made its greatest inroads into religion in the middle and latter half of the nineteenth century. At about that time it became obvious that a sort of credibility gap was opening between what could be explained within the framework of religion and what could be explained within the framework of science. My own view is that this gap is still widening, and widening in a curious and unpredictable way. This is because the more that scientists discover about the universe, the less explicable it becomes. That is a statement which might not have been made at a rationalist meeting thirty years ago—although it is very difficult to give precise dates—because I think that until then people felt that science was gradually explaining the entire universe. They thought that the more scientific facts that emerged, the more we would know. It sounds simple and reasonable enough, but I think that as rationalists and Humanists we should realize that there is a weakness in our own armour here. We can no longer really suggest that within the foreseeable future—within one lifetime or even within generations—that science will be able to say anything fundamental about the true nature of the universe.

To return to the nineteenth century, a development which was enormously significant in religious and scientific terms was the evolution of Spiritualism. Today Spiritualism seems to be absurd and a large part of it seems to be ridiculous, but it becomes less ridiculous if looked at in the context of what people believed and felt in Victorian times. This factor partly explains the enormous interest in Spiritualism on the part of some of the top scientists of that time; in fact, of the key scientists, a substantial number, perhaps even a majority, were interested in Spiritualism and took the trouble to seriously investigate psychical research. This is a very interesting fact which needs some sort of explanation and I think that the explanation is fairly simple.

Everyone in this country and in the western world up until, I suppose, 1880 or 1890 was brought up with a very formal, rather simple-minded, religious training. Men like Alfred Russell Wallace, Conan Doyle, Crookes and so on were brought up to believe certain basic simple facts which are inherent in the Christian ethos; and one of these facts is that souls are immortal. There is another world, or other places, where people live after they die and, depending on whether they are good or bad, they go to one place or another. If one could somehow resurrect Sir Oliver Lodge and say 'Right, what do you feel or what did you feel?', he would say 'I never had any doubt that people survive death'. Now the phenomena of Spiritualism, of course, were nothing more than an attempt to prove what was already believed by the vast majority of people. Spiritualism was an attempt to prove not only that people did survive death, but also that communication with them was possible. In Victorian times this was not an absurd idea at all. If you could set up a telescope and look at a planet, or if you could pick up an instrument and send a message in morse code across the Atlantic, why should you not be able to contact this 'other place' to which people went after they died? Take radio for example: this was a perfectly incomprehensible idea then and it is not much more comprehensible to most people today. There is nothing more unlikely, in principle, in being able to talk across the Atlantic or to Australia than there is in being able to talk to people who are dead, if they continue to live somewhere else.

The Spiritualist argument is really fascinating and I imagine that most people in the audience have had at least a look at it. I think we need to face up to this idea and ask ourselves a question: if Spiritualism is absurd, how can it be explained? Rationalists feel intuitively that it is absurd, as did many Victorian scientists. I think it was Huxley who, when he was invited by a close friend to witness some marvellous materialization, some wonderful evidence, which a Spiritualist medium was providing, said that even if this were true it would not interest him. He would not cross the road to listen to all this rubbish coming from the 'other side'. He was not interested in the subject and anyway it did not matter how many times he saw evidence, he still would not believe in it. That, of course, is a very unscientific attitude to take, but very much the same sort of thing happened recently as far as parapsychology is concerned. J. B. Rhine at the Parapsychology Laboratory at Duke University had been publishing a great deal of data in reasonable reputable scientific journals which, if taken at face value, suggested that ESP had been proved. Yet D. O. Hebb, a psychologist who to my way of thinking has made one of the greatest contributions to modern theoretical psychology, said that he did not care how impressive the evidence was for telepathy, he simply did not believe it and that was that. It just did not fit and it was not good. Now, again, this is a monstrously unscientific point of view, but I think *post hoc* analysis has shown that Huxley and Hebb and all the other doubters have turned out to be correct.

What *was* the data of Spiritualism? Mediums sat in rooms or cabinets and produced ectoplasm and spirit forms and messages from people who had died. They even took photographs, some of which were really remarkable pictures. There is one picture in particular which is critical and represents a make-or-break belief point as far as one's attitude to Spiritualism is concerned. It shows Sir William Crookes, President of the Royal Society, arm-in-arm with the materialized form of a spirit, a girl with a veil, called Katie King. It is not a double exposure, or any other form of contrived picture because Crookes himself said, 'yes, I did have my arm through Katie King's arm and I saw her appear out of the floor, and walked around the room with her. The medium Florence Cook was in the cabinet and I could look in and there she was. It wasn't Florence Cook dressing up as I would have seen if she'd been dressing up.' So we have a whole series of photographs of a President of the Royal Society arm-in-arm with a spirit form.

This is either one of the most important photographs ever taken or else Crookes was either deluded to the point of insanity or guilty of really appalling deception. That is why I say it is a make-or-break. If one could show that a scientist of Crookes' standing was so blatantly deceitful as to pose for photographs with either a woman dressing up or an accomplice, then human testimony on this subject is valueless or very nearly so. How can one take this? Is human testimony about material of this kind valueless or not? If it is, it throws out of the window reams and reams of data along with book after book about psychic experiences and ghosts.

To my way of thinking, one need not examine every case reported. You can still believe that Crookes really was arm-in-arm with a spirit if you like. (As it happens, there is more recent evidence to suggest that he was having an affair with the medium Florence Cook and was actually a party to the whole fraud.) But it seems that one's knowledge of the immense fallibility of human memory and the immense weakness of people's capacity for reporting facts objectively, should be remembered. I do not believe that there is any case for attaching credence to anecdotes or stories of that kind. My view is that towards the end of the nineteenth century the scientists who had once trusted in spiritualistic phenomena suddenly began to disbelieve. Medium after medium was caught in fraud, people were found dressing up and being photographed by flashlight and gradually scientists became disillusioned. Of course, those who had become deeply committed never really changed their minds and they became guilty of acts of credulity which are funny at one level, but tragic at another.

By the end of the century the subject was beginning to lose potency. Scientists pulled out of Spiritualism and, more important, there was also an increasing disbelief in life after death. As soon as people began to doubt the notion of personal immortality, the Spiritualist case began to deteriorate. It has since had two surges, quite predictably, at the end of each of the major wars of this century—major revivals which were due to people's grief at losing members of their families.

The next phase in the saga can be classed under the heading of Psychic Research. (The purist may say that there is little difference between the two, but there is a very big difference in motivation: Spiritualism attempts to prove that people survive death; psychic research demonstrates only that strange things happen—ghosts *may* exist, and weird odd phenomena *do* occur.) This phase lasted from about 1920 to about 1950, when the focus of interest shifted to haunted houses, poltergeist phenomena, talking mongooses and so on. It was a professional hobby for some very intelligent men in the thirties and forties and some remarkable people became involved. They took what they thought was a very hard line: that they were not Spiritualists, did not believe in the survival of the personality after death but did believe that there were strange psychic forces which could be activated by the mind, and that there were disembodied entities of one sort or another. The key figure in this era was Harry Price, a professional ghost-hunter with a large private income, who spent a great deal of money on researching ghosts and wrote many books on his findings. He was best known for putting forward the story of Borley Rectory, one of the most widely-quoted and impressive cases of the era. Indeed, Borley was a remarkable saga and it looked very impressive if you took it at face value though it is now known that the Borley material was spurious and that Price was involved in manufacturing most of the ghosts himself—a very remarkable discovery which was made in the fifties by the Society for Psychical Research. In a way, the discovery that Price was involved in hocus pocus removes him from the field of objective witnesses and signalled the end of that phase of interest in haunted houses.

As psychic research, haunted houses and the like became less and less acceptable in scientific terms, we move into the era of parapsychology: psychical research which involves tests of telepathy. It had its great vogue in the forties and fifties, largely due to one man, Professor J. B. Rhine, a young botanist influenced by McDougal, the Professor of Psychology at Duke University, North Carolina. McDougal believed in the importance of psychic research, and he persuaded Rhine to investigate it scientifically in the laboratory. He also persuaded the University to finance a small department which flourished from the thirties until the mid-sixties, when Rhine retired. The evidence for telepathy during the forties really looked good, both to psychologists and to scientists. And yet experiments were criticised as being carelessly done, with none of the rigour that was necessary, so that the explanation that the information supposedly gained telpathically was coming across by normal means was quite sufficient to discredit it.

Having summarized scientific interest in the topic, there remains the aspect of 'lay' interest. As scientific work has undermined religious beliefs, it has left an enormous gap. Whatever the achievements of science in increasing information and in practical invention, it still does not answer the 'cosmic-significance' questions that people continually ask, and the conceptual gap still widens. Very few people have a proper scientific education, many are partly educated in science and only become aware of how complicated the problems really are and that science does not have all the answers. So, in various ways, people have started to relate themselves to the universe, the unexplained part of the universe which traditionally has been the province of religion. All the cults that have interested me are still concerned basically with survival, with the idea of human personality as perpetual in one form or another. They do not necessarily take the old Spiritualist/Christian view of life after death which is 'situated' in 'another world', but they do imply a form of continuing existence. Each of these cults is interesting inasmuch as there is a quasi-scientific theme running through them. They pick on some of the important concepts of science and psychology with which people are familiar, and mould themselves around these ideas.

The most interesting of these cults or modern religions is Scientology—it is certainly the most interesting *organized* body. The cult began as 'Dianetics' over twenty years ago, founded by L. Ron Hubbard, a successful professional writer not only of science fiction but of westerns and true romance, detective stories and fantasy of all kinds, and also of many articles about such subjects as aviation and boating. He is an amazingly interesting man, an entertaining conversationalist and has the charisma that is very important for popular leaders.

In the late 1940s Hubbard conceived and began to develop ideas about a special kind of psychotherapy. He thought that he had some kind of insight into the nature of the human mind, which he envisaged as a very simple thing run on simple principles. Once one grasped these simple principles there would be no problem in using them to heal people whose minds had gone wrong and to make ordinary minds supernormal. Built into the theme of Dianetics, one can discern concepts drawn from two very important literary fields: one is in science fiction, a *genre* which enjoyed wide popularity in the forties and fifties, and the other is psychoanalysis, also at its peak at that time. (The fifties were the last great years of psychoanalysis in its old form, when the principles were becoming known to a very wide public through paperback books.)

The thesis of Dianetics was published in *Astounding Science Fiction*, the most successful science fiction magazine with a very wide circulation and an indisputably literate readership. It was published with an enthusiastic endorsement by the editor as being one of the most important articles that had ever been published—a statement typical of Scientology. It fell on exactly the right ground. Science fiction fans are very receptive and open-minded people, always ready for anything like this; they welcomed Dianetics. In simple terms, its principles were as follows:

A man has two minds, the conscious mind and the unconscious mind, rather as in the old Freudian sense. The unconscious mind can gather data—particularly unpleasant incidents that have happened to one—which attach themselves to it and handicap it. Such incidents also, *ipso facto*, handicap the individual, who is unable to understand what is happening. It is an old idea, but Dianetics presented it in a fresh form and new language. The really important point was that if one can only be brought to identify this 'engram' (the point at which a past unpleasant incident shackles the unconscious mind), to seek and confront it, it will then vanish and lose its power. What Hubbard said was that whereas psychoanalysis leads only slowly to a confrontation with the traumatic event, Dianetics leads to it quickly. A matter of two hours or so of Dianetic therapy could bring a man face to face with his engram, and then it would disappear. Furthermore, what was particularly enticing was that a knowledge of medicine or psychology was unnecessary—one simply had to know the rituals and techniques of Dianetics to be able to do this oneself, and in no time at all obtain results which were much better than any that a psychologist or doctor could produce.

Within weeks of publication of the article, Dianetic groups were spreading throughout America and even through Europe. People were having Dianetic processing and were praising it; they claimed that they had never felt better, that their ailments had disappeared.

The next thing that Hubbard discovered was that some of these engrams were induced during pregnancy while the foetus was in the uterus. If something unpleasant were done by the father to the mother, such as kicking her in the stomach while she was pregnant, the engram would be stored and the memory of this horrible event would shackle the poor foetus for the rest of its life. Hubbard postulated that the essential element of human life is a conscious centre, an *ego*, a personality which he called the 'Thetan', which is fully conscious from the moment of conception and even further back, and therefore could be aware at one level of strange and unpleasant prenatal events.

In time, many people became disillusioned with Dianetics because it did not seem to have any staying power, but interest revived when Hubbard discovered that the engrams could even be laid down in a *past* life. The Thetan was in fact a perpetual entity existing throughout the whole of time, and inhabited body after body to endless reincarnations. Therefore, of course, people did not have only one lifetime's engrams to contend with, but also many more from past lives. So no wonder, said Hubbard, that some people were unhappy. Dianetic processing enabled people to recall past events, past lives and past deaths, and they would remember various things that had traumatized them, a factor which is still a very important part of Scientological processing today.

The difference between Scientology and Dianetics seems to be small. Dianetics was taken over, in the 1950s, by a group of wealthy businessmen who were very impressed with its effect on them but who did not think that Hubbard should be in charge of something like this even if it was his own invention. So they formed and financed a Dianetics Corporation, acquired the rights to Dianetics and its books and set up a Dianetics Research Foundation in Wichita. But Hubbard, of course, was far too independent a character to be tied down to a couple of oil millionaires from Wichita, so he soon withdrew and founded Scientology, in Phoenix, Arizona. Without Hubbard, the original Dianetic movement was absolutely nothing and it petered away in a year or two. Now Dianetics is back again in harness with Scientology theory.

When Dianetics went into a decline in the fifties, Scientology never really got off the ground, but it revived again recently and has changed its form very dramatically in the last five or six years. For anyone interested in the evolution of religions, the history and sociology of religion, it really does merit a close investigation. It has evolved so much more quickly than religions a thousand years ago or even five hundred years ago could have done, because, of course, the modern media can propagate information so rapidly that ideas can spread in a matter of years instead of in centuries. Scientology went through a very repressive, paranoid stage when it was under attack (justifiable paranoia perhaps) and it has gradually worked round to being a fairly explicit religion; there are religious services with prayers and people in vestments, and they have clergy with dog-collars and all the trappings of religion. The move into becoming a religion was a relatively recent one, and it is significant that Hubbard remarks that a religion cannot be attacked by parliaments and similar bodies. But however Scientology started life, the people who are deeply involved in it, sincerely hold it to be a religion : they treat it as a religion and it has the function of a religion, but a religion without the out-of-date concepts of angels, devils, heaven and hell. Scientology holds that people are immortal, and reincarnate for millions of years. Their hell, if you like, is realized by accumulating too many engrams and consequently having a very difficult time on earth, lifetime after lifetime. The pathway to 'heaven' is through Scientology, which roots out these engrams and restores the superior being to his proper status.

Scientologists have a very interesting gadgetry to help them—the E-meter, a modification of the psychogalvanometer which is a typical piece of very inefficient psychological equipment which measures the change of the electrical conductivity of the skin. A subject aroused or in an emotional state, when attached to a psychogalvanometer circuit, produces a swing of an ammeter needle indicating that the skin resistance has changed. It is a useless gadget really, and is rarely used in psychology, except in laboratory demonstrations, but Scientologists have modified the psychogalvanometer to a couple of cans (soup cans with wires joined to the amplifier and the meter). The person goes through a series of drills that are part of the technical processing, and at the point when the auditor, as the therapist is called, suspects that some sort of engram is present, the needle moves. However, one can induce these needle movements simply by increasing the pressure on the cans, which increases the contact between the skin and the cans and therefore changes the electrical conductivity of the circuit. One can also move the needle by sweat on the hands, because contact between skin and can increases, and thus the conductivity is improved again. So the probability is that the E-meter is testing only people's immediate reaction to the questions that are being asked them. Progress through Scientology (a very expensive business, costing hundreds of pounds to move up from grade to grade and literally thousands to reach the top) is monitored by ability to handle this meter, and one suspects, of course, that people get very skilled unconsciously at making the meter do what they want.

If Scientology is indeed evolving into a religion, one would expect it to offer an explanation of the universe, because all other religions do. In fact, it does offer a very interesting 'cosmic plan' which Hubbard has written about in his books. His thesis is that people are decadent gods. The universe originally consisted of a number of beings—Thetans, as he calls them—who were omnipotent and omniscient, with all the other attributes of gods. The trouble was that there were only Thetans, and as they were all omnipotent and omniscient, their world was a dull place. So, to while away the time, the Thetans began to play games of various kinds. But a game is boring if you know the whole sequence of events in advance (which is what happens if you are omnipotent) so they started deliberately to handicap themselves. They did this by creating physical universes with mountains and seas and so on, and they would become involved in these universes, themselves living in bodies. Gradually the Thetans found there was a trap here: they were enjoying the games so much that they began to forget that they were only playing games and gradually they became embroiled, until here *we* are and we have completely forgotten about the origins of the thing. It is fortunate that L. Ron Hubbard has explained to us how we got ourselves into trouble all those millions of years ago.

To recap : for a modern religion to have any credence, to *do* anything for people, it must play with the concepts of the time. Archangels and cherubs are dead and no one believes in them now. Belief in flying saucers exemplifies this point. There is no *one* predominating religion involved in flying saucers, but there are a number of minor groups—some of them growing quite fast—which use the ideas of flying saucers and space travel as a framework to build the house of their religion. In point of fact, one can make out quite a good case for the existence of strange objects flying around in the sky.

The Condon report on UFOs, commissioned by the United States Air Force at the University of Boulder in Colorado, maintains that there is really very little evidence that they exist, but produces a number of cases which have no apparent natural explanation. There are one or two cases in which fighter pilots have tracked things moving through the sky and have seen them ahead, things which have also been plotted simultaneously on ground radar; in other words, the object the pilot was chasing was visible to him, and both his plane and object were tracked on radar. Several other cases are admitted to be inexplicable. The report merely states that there is not enough evidence for any recommendation that the Airforce spend any more money trying to research such cases (which is, of course, the key thing).

To take a slightly broader view of things, one sees that it is rather ridiculous to believe that man is the only intelligent form of life in the universe. The balanced scientific view at the moment is surely that there probably is intelligent alien life, and that there may be a good deal of it in the universe. Of course, there is almost certainly no other such intelligent life in our solar system, but there could well be some in other planetary systems, a matter of light years away. These distances are enormous so we tend to dismiss the idea of travel across them. We are in a relatively young part of the universe and if man has progressed to space travel in some three thousand years of recorded civilized history, alien societies, which might have a million years of history behind them (assuming that they do not destroy themselves in the course of evolving), might be expected to have made quite substantial advances in long distance space travel. So it is not unreasonable to take the view that aliens might be moving around the universe in space ships, with an interest akin to our own and similar aims of research and discovery.

Few people would really disagree with that point of view. But the saucer believers go a stage further and state that unexplained objects in the skies—the UFOs—*are* space ships and that Earth *is* being visited by alien life forms.

Flying saucers, or flying oddments to be more exact, apparently started landing on Earth in large numbers in the early 1950s, and since then, more and more people have reported not only seeing them, but also making personal contact with them. The first, or perhaps one should say the most widely publicised of these, and sensational at the time, was reported in the book *Flying Saucers have Landed*, by Desmond Leslie and George Adamski. Today, it is difficult to understand how anyone could take this book seriously. It is a ludicrous and unimaginative account of a meeting in the desert between

Mr Adamski and a long-haired Venusian, who gave him a solemn warning (communicating in telepathy of course) that the world was in danger from its own nuclear experiments, and would incur the wrath of the normally peaceful Venusians if such experiments did not cease. The book is too childish and trivial to warrant much attention, except as an exercise in the study of human credibility; it is interesting that it was swallowed with such enthusiasm by the world at large and is still referred to with awe by many people today.

The question we must really ask about flying saucers is this: Why do people rush with such eagerness to believe in them? The great psychologist Jung was once interviewed about UFOs in general and made a balanced and tolerant remark to the effect that it was conceivable that intelligent life exists elsewhere in the universe, and that such beings might have developed space travel. This modest and reasonable statement was immediately interpreted in newspaper headlines as 'Jung believes in flying saucers'. This distortion interested him a good deal, as did the fact that his own subsequent statement to the Press, stating that he did *not* believe in saucers, was totally ignored. This raised in his mind the question as to why the news was published that saucers existed, and therefore that such news was presumably welcome, while news that they did not exist was apparently unwelcome. Jung finally came round to the bizarre conclusion that the UFOs were indeed real, as the psychic projections of people's minds—projected wish-fulfilment images in the skies. While I do not believe that UFOs have any kind of external reality (other than when they are misidentified natural objects), I do nevertheless feel that since Jung wrote his book, the evidence that they reflect deep and growing anxieties within mankind has been building up steadily. People seem to be realizing, with increasing clarity, that the old gods who have proven such reliable father-figures in centuries past are vanishing into the mists of incredibility and have left no obvious surrogates behind. No longer are there any super-beings to turn to when individuals or the world itself gets into trouble—at least, none of the old-established variety. But here of course lies the real attraction of flying saucers. Suppose (and it is not unreasonable to do so) that the universe is populated here and there with intelligent life, then it could well be that civilization hundreds, thousands and even millions of years in advance of our own do exist. It is difficult not to believe that these will have evolved space travel, and very sophisticated space travel at that, and could very well be capable of visiting Earth. Whether they *could* do so is another matter, and whether they would make themselves known to us even if they did visit Earth is also very uncertain—reflect on the dire consequences of mixing advanced and primitive cultures on Earth for example. Nevertheless, Super Space Beings are infinitely more credible as potential rescuers of mankind than the gods and archangels of the past, and they represent the elements of an eminently suitable space-age mythology. Rationalists and Humanists a quarter of a century ago must have felt that they had well and truly routed the forces of superstitious religion. And so they had, but how ironic that new gods and myths should so speedily be created, and that they should be cast in the image of modern science and technology! Perhaps this, coupled with the growing interest in other quasi-psychological cults, such as Scientology, should warn us that religious beliefs, whatever form they might take, still play a very significant rôle in the philosophy of man, and may do so for far longer into the future than nineteenth century rationalists would have predicted or cared to believe.

COLIN CAMPBELL

A Rational Approach to Secularization

THE TASK WHICH I HAVE SET MYSELF in this paper is essentially an exercise in de-mythologizing. As I have always regarded one of the principal functions of rationalism to be the elimination of false belief, it follows that a rational approach to social problems must involve an attack on the myths concerning the nature of the society in which we live. That we all do cling to myths about the nature of our own social world is as certain as the fact that we all cherish false beliefs about ourselves. Indeed, the erroneousness of the beliefs stems from the same cause—our desire to believe things which are in our interests or which rebound to our glorification. It is for this reason that I have chosen to discuss the myth of secularization.

I have already betrayed my position by saying that I regard secularization as a myth, a position which I shall try to justify. It is obviously in our interests to believe that secularization is occurring. To hold such a belief is to have the comfort of knowing that history is on one's own side and one is clearly not going to relinquish that belief in a hurry. Interestingly enough, however, it is also the case that nearly everyone else in society, including the clergy, also believes that secularization is occurring. So here we have an example of a myth so prevalent and persuasive that it is held even by those who have no interest in believing it.

What I shall attempt to do, in a discursive fashion, is to throw out ideas surrounding the problem of deciding what secularization really means and what information we possess and how we can interpret it. I shall suggest problems, point out difficulties of interpretation and identify misconceptions rather than strive to reach any clear-cut conclusion. For this is an area in which, the more you inquire into it the more you discover the very considerable difficulties of interpretation surrounding the little information available.

The obvious place to start is at the general position of an increasingly secularized society that most people accept. Most people would say that it is obvious that secularization is occurring in contemporary society because there is a decline, proportionately and absolutely, in church membership and church attendance. In Britain this is represented in most discussions of the subject by an apparently unambiguous graph covering the period 1900–1950 and showing a steep decline in church membership of all the Protestant denominations, with a particularly dramatic dive in the case of Sunday school membership. Of course, as soon as one does this exercise, one realizes that there are some exceptions to be made to the rule. The Roman Catholic Church is represented by a slightly upward-sloping gradient attributable in large measure to immigration from Ireland in the early part of the century. Also, the small sects, like Scientology and Jehovah's Witnesses do not display a decline but have shown slight increases in membership. Nevertheless, despite these exceptions, a general downward trend is discernible. Thus, there is statistical evidence of an absolute and a proportionate decline in the number of people who are members of religious denominations as well as in the number of people who attend Sunday services.

I could continue at some length about what these statistics actually mean, given the way in which they're collected. One thing that sociologists are very prone to do is to discuss the reliability and validity of statistics. Most social statistics have a very low validity and reliability, and this is certainly true of religious statistics. To take a few examples: some religious organizations do not regard women as members and therefore do not include them in their returns. This is a form of sex discrimination that you may, quite reasonably, object to. Some denominations count children and some do not. Most of them collect their statistics via clergymen themselves who are not usually the best lay sociologists. So there is a whole host of reasons for being doubtful about these statistics. But the trend appears to be so straightforward that no one would wish to dispute that a decline is indicated. However, it is somewhat suspicious that all observers use the same particular time period; they all refer to the period from the turn of the century to the 1950s. It is well known that one of the best ways to fiddle statistics is to use an arbitrary time span. If, in fact, we were to go back in time so that we covered a complete one-hundred-year period from 1850 to 1950 things might look a little different. Of course, part of the problem here is that we are getting into regions of very doubtful statistics indeed. But the one and only religious census in this country was conducted in 1851 and in spite of the difficulties in interpreting these figures they do supply some information about what the situation was like at mid-century. In addition, there are other statistics of various sorts to provide a clue to what might have happened between 1850 and 1900 and the general indication is the reverse of the twentieth-century trend. In other words, one could suggest that there was something like a boom in religious membership and church attendance in the second half of the nineteenth century, the 1870s and 1880s in particular being a time of intense religious activity.

Looked at in a broad historical perspective, it is as interesting to know why people were so preoccupied with religion towards the end of the century as to know why they appear to be so uninterested in it now. Perhaps there is a level or norm of interest in religion which we may not be very far off at this present moment, and what we have witnessed over the last fifty or sixty years has been a boom which petered out. Although I do not wish to press this interpretation, it does seem as plausible as arguments which treat only the period from 1900. There is a more general and highly speculative question concerning the degree of interest people had in religion in earlier historical periods and whether or not there is evidence of a cyclical pattern of interest in religion which alternates with interest in political affairs. Certainly the decline in interest in religion in this country just after the turn of the century was related to an increasing concern with political affairs and in particular with the rise of the Labour Party.

There are two views of the nature of secularization which are roughly analogous to the contrasting themes of the origin of the universe. The most popular view is what one might call the 'big bang' theory of secularization. This is the view that considers that once upon a time, way back in the mists of pre-history, there was a form of human society that was completely and totally religious. In every way conceivable and in every conceivable activity, behaviour was permeated with a lingering religiousness. However, ever since that distant time, religiosity has been running down. In accordance with some spiritualized law of entropy, the religious element in life has been draining away. Such a 'big bang' theory necessarily suggests that the further back in time one goes, the more religious societies become. It also intimates that an inexorable process is occurring which will eventually mean that all religion will drain away from social life, leaving a totally secular position. Opposed to the 'big bang' theory is the much less popular 'steady state' theory of religiousness. Here the assumption is that, in place of an original explosion of religiosity, there is a continuous creation of religious feeling and concern occurring in the human situation; religiosity is being created at least as fast as it is being consumed through the trials and tribulations of mundane secular existence; and that in consequence there is really no great fluctuation in the amount of religious concern in society over time. These, then, constitute the two dominant theories concerning the nature of secularization.

In all probability most people hold to a modified version of the 'big bang' theory. The main problem here is to know precisely what interpretation to put upon this theory. One possibility would be to assume that secularization means disenchantment or 'desacralization' of the world and thus that the further back in time one went the more one would find spirits inhabiting every bush and tree and every spring and every rock; gradually, over time, such spirit entities were displaced by empirical analysis. The supernatural is pushed back and up into the sky, first of all as a pantheon of gods displace spirits and then as monotheism displaces the pantheon. There are certain problems with this general argument. Firstly, such a view would grant Judaism the status of one of the best secularizing agencies that the world has seen. Monotheistic religions in general are principally concerned in this very sense with the destruction of a diffuse magico-spiritual world. Yet according to the theory such a process is secularization. This ambiguity highlights one of the central problems involved in any exposition of secularization theory, that is, the difficulty of distinguishing secularization from religious change. Secondly, assuming that one can demonstrate that the quantity of the sacred is actually diminishing in its traditional locations, one can still not be sure that the sum total of society's sacredness has diminished. Some sociologists would argue that the sacred is driven out of one resting place only to become embedded somewhere else but that we fail to see this because we are so accustomed to looking for manifestations of sacredness in its 'traditional' resting place. They would claim, for example, that, in our own society, notions of sacredness have become embedded in the ideal of romantic love and in pop culture and even in sport. This is a view implicit in the 'steady state' theory.

Other problems surround the positive side of the 'big bang' theory, for example, the assumption that if the sacred is diminishing then the rational empirical content of man's thinking will be increasing. Yet there is not very much evidence that if people cease to view the world in terms of sacred entities they necessarily turn to a rational–empirical view as a replacement, or that they become more 'scientific' when they lose their magical beliefs. Indeed, one could argue that all that happens is that traditional beliefs become replaced by various new forms of superstitions or 'subterranean theologies'.

There is, however, a different argument one could advance in defence of the conventional secularization thesis. For, surely, if we are in some doubt as to what is happening to the beliefs people hold, there is no doubt that the influence of organized religion has declined. The institutions of religion are obviously and visibly in a state of decline. But this assumption too is much less self-evident than it might appear. It is true that there has been a decrease in the general level of involvement in traditional forms of religious expression. But then it should not be surprising that traditional religious forms tend to go out of favour in a rapidly changing society; it is not a very startling conclusion to come to, that something traditional is in decline in a changing situation. One could equally show that traditional forms of the family are in decline, and that traditional forms of economic activity are in decline. In other words, whatever is defined as traditional is necessarily in a state of decline in a rapidly changing context. Clearly, that kind of observation proves very little and yet this is frequently overlooked by those who seek to measure the rate of secularization in our society.

Frequently religion is implicitly defined as that very specific form of evangelical Protestant Christianity which existed for a comparatively short period in British society, and then indices are devised to show how far contemporary populations have 'lapsed' from such a faith. One poses statements of belief, such as how far people accept the divine creator God; how far they accept the notion of an after-life with a distinct system of rewards and punishments; how far they accept the notion of a literal interpretation of the Bible—beliefs which formed the cornerstone of *Victorian* religion—and then one tries to see how far people today hold to such beliefs. Although it may be interesting to know that such a change has occurred, it is not very relevant to label it 'secularization'. One could, on this basis, take any period of history and, showing that there was a change in the nature and content of religious belief, dub that secularization. This is, in fact, precisely what some American sociologists have done and, having devised a scale of orthodox belief, they set out to discover which groups in society were the most religious and which the most secular. It is interesting that their results indicated that one of the most secular groups were clergymen. Thus one scale of secularization suggested that the most secular people were clergymen and that their congregations were much more religious than they were. This in itself should suggest some cause for a reappraisal of this kind of exercise such that full allowance is made for the distinction between religious change and secularization. If someone has a very strong commitment to the new theology, he will automatically be dubbed as secular on all these scales. It is really as silly as saying that since one does not see very many hansom cabs around these days therefore transport is in decline. The whole exercise is quite futile unless one sets out some notion of the general nature of the religious response in society and makes some allowance for the emergence of new forms of religious response, in addition to measuring how far the conventional forms have gone into decline.

Clearly, what is required are two scales, the one measuring religiosity and secularism, the other measuring traditional and modern forms of religiosity, but as yet such scales have not been devised. One has to combat the natural tendency to think that religion reaches some kind of apotheosis of truth at a particular moment of time and that, therefore, any subsequent change is secularization. For evidence would indicate that religious traditions continually change and become re-interpreted in order to meet the needs of their own time and that this change is continually being confused with the process of secularization. In any attempt to assess secularization there is thus a need to be aware of the extent to which new religious institutions are emerging. The fact that people only infrequently go to church on a Sunday may be simply an indication that the cult of the Sunday sermon has been replaced by new habits. If, in fact, one looks for evidence of the emergence of new forms of religious institutions, one can find it. The emergence of the 'house church' is one example of informal religious organizations set outside the context of the church itself. Indeed, as soon as one starts looking for evidence of religious activity outside the formal structures of the church, one will find sufficient data to suggest that much of the apparent religious decline may simply indicate the extent to which people have become disenchanted with the traditional methods of expressing religiosity rather than with religiosity itself. Additional evidence for this can be found outside the narrow framework of Christianity. Indeed, the conventional thesis of secularization retains its credibility only because it is so frequently interpreted to mean 'de-Christianization'. In general the other world religions have not experienced the decline which has hit Christianity and, although they constitute only a very small part of the total picture of religious life in Britain, their experience of success is significant. Interest in Hinduism and Buddhism has definitely increased considerably since the 1940s. Islam is also gaining in adherents but largely as a result of immigration. The point is that the argument that religion in general is in decline just will not agree with the evidence of the increasing interest in other world religions. Also, it is the case that interest in religion in general shows no evidence of falling away.

However, having conceded many of these points, one could still argue that secularization—conceived as a decline in the power of organized religion—is occurring. Certainly, religious institutions are less directly involved with other spheres of social life than they used to be. The Churches exercise much less control over such fields of life as education and social welfare than they did in the last century and earlier. In this sense, we are noting a process involving the loss of functions that is paralleled in other social institutions. The family is also less involved in providing education, social welfare and economic activity than it was in the last century. Whether one would necessarily conclude that the family has lost influence is debatable, for one could say that this has meant that the family is now more able to concentrate on those functions which are unique to it. In other words, because we do not have to bake all our own bread and we do not have to make all our own clothes within the family, there is more time for adults to devote to initial socialization of the children. Exactly the same argument can be advanced concerning the Churches, that is, that the loss of 'extra-religious' secular functions means that the Church is now more able to concentrate on its distinctive and special concern with saving souls. The loss of these other functions is thus to the benefit of religious institutions because it enables them to pursue their own distinctive purposes unencumbered by involvement in secular activity. Here the notion of secularization is being given a meaning at odds with its traditional usage. For example, if one considers what happens to a religious sect which refuses to be involved in society in an attempt to retain its circumscribed religious purity, and then subsequently becomes successful and powerful, spreading through society and involving itself in secular activities, one could, in the above terms, talk about that process as secularization. It is not, however, secularization of society but secularization of a particular religious group. Consequently, one can argue that the fact that the Churches are becoming less involved in extra-religious spheres of life is a process of desecularization. Indeed, some Christians take comfort from this argument and claim that this is going to enable the Christian Churches to take up their old traditional prophetic rôle on the margins of society; that no longer being near the centres of power is exactly what is required to enable them to be true to the intrinsically radical nature of the Christian message, and that they will benefit from not being compromised by their involvement in secular activities.

Clearly, the argument concerning the degree of influence that religion is able to exert in a society is a complex one, but it is plausible to argue that religious organizations can exert more influence by not being directly involved in the centres of power than by being involved but compromised by that involvement. There is another aspect to the argument about influence which is relevant here and this is the debate over whether religion exerts its influence in society primarily through organizations or through the individuals that they comprise. While we are accustomed to thinking of Christianity as exerting its influence principally through organizations like the Catholic Church or the Church of England, there is another pattern whereby people exert their influence as individuals rather than through organizations. Frequently one may encounter the argument that religion is more likely to be influential in society if people hold firmly to religious views and practise them conscientiously in their everyday life rather than if they concentrate on pressing for particular reforms or legislative measures through the organizations they belong to. (One will find the same debate taking place within non-religious bodies like the British Humanist Association.) Essentially, the debate is whether exemplary or ethical prophecy is the best means of changing people's lives. Are you more likely to get converts if you live an exemplary life and others become influenced by the ideals they see displayed in the way that you act, or if you try to tell them what they should do and encourage them to band together to change the structure of society? This is a long-standing dilemma, whether to try to change the social structure to achieve your ideals, leaving individuals untouched, or to try to change the individuals, leaving the social structure untouched. Religions in general have been in tune more with the latter than the former path.

I have been arguing up to now that the case for secularization is unproven because the distinction between secularization and religious change is rarely made. And this brings me into the very difficult and unresolved area of definitions. What do we really mean by the terms religious and secular? One thing that is generally implied in the use of the terms is a notion that they are necessarily exclusive—exclusive, that is, in the sense that an increase in the one necessarily implies a decrease in the other. How can one so define these concepts that they are mutually exclusive? There are certain pairs of opposites one can use, like theistic and non-theistic, or supernatural and non-supernatural, which are necessarily exclusive but do not exhaust the meaning of either 'the religious' or 'the secular'. It's very difficult, in fact, to define these terms in such a way that they are genuinely exclusive of one another and still cover their general meaning. There are many possible meanings of 'the secular' such as the utilitarian, the mundane, the practical, the rational and the hedonistic. However, many of these concepts have happily 'co-existed within' or even been promoted by one religious tradition or another, so one is still left with the problem of trying to find some way of defining the secular and the religious so that they do exclude one another. But if one does not make this assumption, then a whole new host of possibilities opens up. For example, secularization and religious revival might actually proceed hand in hand. Indeed, some people have argued that there have been periods in history when there has been a considerable upsurge in secular activity and thought, together with religious activity and thought, and periods when they have both been in decline. So it seems as if the general assumption that the religious and the secular are exclusive terms is questionable. The real problem is actually that of defining the secular and not the generally recognized problem of defining religion.

We are still left with the problem of deciding what to make of the term religion and of how to use it. I have already said that we cannot define it in terms of a specific orthodoxy. We cannot define religion as what the Victorians believed or what people happened to believe in 1550 or 1300. Any definition of that type is necessarily useless for this kind of exercise. We want some general notion of what religion means. Unfortunately, any substantive definition of religion is open to problems of ethno-centric bias. As soon as one starts to define religion in terms of its substantive content, one comes across this difficulty. If one defines it in terms of some particular notion of the supernatural, one falls into the trap of excluding some phenomenon which at some time or other has been regarded as distinctly religious. The best example of this is in Hinduism, where any shade of theism, atheism, agnosticism or polytheism can exist, provided that one accepts a common set of traditions and practices. So one immediately comes up against the problem that if one starts defining religion substantively one introduces an ethno-centric bias so that one is merely saying that the secular is that which is most different from the religion of the particular cultural epoch or period of history which one has chosen.

There does not seem to be any successful resolution of this problem. So many people have turned instead to attempts to define religion functionally rather than in terms of its substantive content. Definitions in terms of belief are misleading anyway because religious phenomena consist of belief, practice, experience and organization and so a definition in terms of only one of these is in fact already biased. To assume that religion is essentially connected with belief is to betray a rationalistic bias. This is probably the bias we are most guilty of. In primitive societies religion is based on ritual and practice and only in a very secondary sense is there a concern with belief. But even in our own society there are many people, who, when asked to define religion, would not define it in terms of belief at all, but would define it in terms of practice, like church-going or prayer. Or they would define it in terms of morality rather than belief. And we really have no right to turn round to them and say 'I'm sorry but we will not accept that definition of religion. You will have to define it in terms of belief.' The various traditions in the definition of religion have respectable pedigrees and they will continue to exist, one assumes, for a long time to come. Thus if one is going to use a substantive definition of religion one ought to define it in terms of belief, practice, experience and organization.

One of the things that might be happening to religion in our society is that its various parts are breaking up, they are no longer existing in a clear, complex, integrated whole of orthodoxy anymore, which, if it's true, is making life even more difficult for us. Another way of defining religion is in terms of function, and this is the really popular way of doing it at the moment. We are not very successful in defining religion in terms of a belief in something or as a set of practices, so we define it as that which does something. It is defined in this sense as a system of beliefs and practices which helps to relate man to his ultimate problems, to integrate the social group, to give support in situations of uncertainty, to support his morality, and so on. Such functional definitions have a good intellectual pedigree but they bring a new set of problems. If we define all four-legged animals that run, as dogs, then we are in trouble. Similarly, if we claim that everything that integrates man into the group, helps him deal with problems of uncertainty, gives a faith beyond himself, deals with problems of ultimate significance is 'religion', then we have committed the most elementary of logical fallacies. Many sociologists seem not the least deterred by this, but are quite happy to describe communism, nationalism, fascism, 'psychoanalytism' and all the other 'isms' that they put on their list (including of course Humanism and rationalism) as religions—although they sometimes become a little unhappy when they find that they are forced to include football, pop-music and the ideal of romantic love (for many of these smaller ideologies and movements in society also fulfil some of the same functions). It is therefore necessary to use a functionalist's definition of religion in conjunction with the notion of functional alternatives: the notion of some other system of belief and practice which will fulfil the same function.

If one looks at the literature of the rationalist and Humanist movements in this country, one will find that one of the dominant controversies has always been between two groups that could be called the 'eliminationists' and the 'substitutionists'. The eliminationists have always argued that the principal object is to eliminate religion, that it is something we do not want and that although it may have been useful once (like the appendix), it no longer serves any useful purpose. Bradlaugh would be the best example of an eliminationist. On the other hand, there has been another equally strong tradition which has argued that we will not get rid of anything until it is replaced with something else, because it is fulfilling some useful functions for individuals or society. The argument for trying to substitute for religion is that people have needs which will persist even after the original social institutions are removed and that clearly such needs have to be satisfied.

The literature of the nineteenth century indicates that there was a very vigorous debate between eliminationists and substitutionists about the extent to which people really did have needs which one had to substitute for in a religious sense. The evidence suggests that both were right in that sometimes the needs are carried forward and sometimes they are extinguished. For in some cases the needs die along with the religions while in others the needs continue. For example, the need to feel some kind of continuing personal immortality is a need which religion fulfils for some people. Ethicism provided a way of satisfying this need through vicarious immortality, through the notion of the ethical ideal and ethical action. Some people obviously were glad to have that need fulfilled though they had lost their religious faith; other people lost the need along with their faith. It is really a question of whether, when one suffers a bereavement, one goes on needing somebody else to fill the gap or whether after a time one comes to terms with one's sense of loss and no longer has the need. And, to go one stage further than this, there is plenty of evidence that the needs that people have are the products of their culture, and as religion is part of that culture one can say that the needs people have are created by the religion to which they adhere.

At the moment it is very fashionable to talk in terms of religion as something to do with the ultimate.

Contemporary theology exemplifies this, but it is also the case in sociology that increasingly religion is defined in terms of something to do with the ultimate. Religion is to do with ultimate concern, ultimate significance, ultimate problems, ultimate solutions, ultimate whatever, but it is ultimate. This, of course, stems very much from Tillichian tradition and I think that sociologists have made a mistake in turning to him for help in dealing with problems concerning the definition of religion. As it happens, theologians are also turning to sociology for help and the situation is becoming so confused that one can never really be sure which hat people have on when they are writing in this area—one now described as socio-theology. Sociologists are trying to use the notion of the ultimate as an escape from the definitional dilemma. Substantive definitions of God encounter the problem of historical and cultural bias while traditional functional definitions founder on the problem of functional alternatives. Increasingly, therefore, sociologists turn to the notion of the ultimate and talk of religion as being concerned with ultimate problems of human life. But, unfortunately, they have still not escaped from the cultural dilemma because there are no given ultimate problems. If one looks at the list of ultimate problems presented, they bear a remarkable resemblance to the traditional concerns of Christianity: death, suffering, illness, disease and so on. These are not necessarily the ultimate problems, for what are defined as problems are cultural variables. One cannot make a list of ultimate problems confronting everyone because what is a problem is itself a variable.

The complexity surrounding the notion of secularization and the nature of religious change is such that one should be very, very careful before employing the words secular and secularization in describing our society. One certainly is not justified in making the glib assumption that secularization is occurring. If, on the other hand, one looks at the broader question of the changing relationship of religion and society, free from the constraints of assumed secularization, one may gain a better understanding of what is occurring. There is, however, one sense in which one may still justifiably use the notion of secularization and this is by returning to the theme of myths. Whether secularization is occurring in society or not, it is a fact that people believe that it is, and, in many ways, secularization as a myth is more significant than secularization as a fact. For not only Churches, but also rationalists are busy reacting to secularization in such a way as to make one wonder if this might not be an instance of a self-fulfilling prophecy in the making. However, the main concern is to identify where the rationalist should stand on this issue, and surely his duty it clear, for is it not part of the duty of the rationalist to expose myths even if they are to his advantage? It is an important part of a rationalist's duty to clear himself of myths and to see society and to see himself in as clear a light as possible, as well as to expose the myths of others. If the aim of rationalism is merely to attack and demolish the myths that we and others hold then there is more than enough work to keep us busy for a very long time to come.

should [illegible] what [illegible] to the world [illegible] [illegible] [illegible] [illegible] [illegible] [illegible] the [illegible] [illegible] [illegible] [illegible] [illegible] [illegible] [illegible] [illegible] [illegible] of the [illegible] [illegible] [illegible] [illegible] [illegible] [illegible] [illegible] [illegible] [illegible] [illegible] [illegible] [illegible] [illegible] [illegible] [illegible] [illegible] [illegible] [illegible] may [illegible] [illegible] [illegible] [illegible] [illegible] [illegible] [illegible] [illegible] [illegible] [illegible] [illegible] and the [illegible] but not only [illegible]

ERNEST H. HUTTEN

The Future of Science: Mechanization or Humanization?

WHICH WAY IS SCIENCE GOING? I do not mean by this question to enquire about possible new theories or discoveries. I refer to the *metascientific* enquiry concerning the kind of explanation we expect to obtain from science. Until today the very power of science has resided in the method first formulated by physics. Physical science has set the paradigm of a scientific explanation and established certain standards of objectivity and of reality. Causality and atomism are the two main conceptions that have brought about what we call now the 'mechanization of the world picture'. But physics itself has changed radically and, in fact, has given up the model of a mechanistic universe. The human or social or life sciences have never fitted into the mechanistic scheme of explanation. And although many attempts were made to shrink, say, psychology to fit the Procrustean bed of physics, by cutting off parts of the living organism, none of these attempts have succeeded in finding general acceptance. Thus the question arises whether modern, non-mechanistic physics and the human sciences cannot come together and develop a new and more adequate method of explanation. What is the difference between mechanistic and modern physics and what new approaches do the human sciences, for instance, psychology, offer to our understanding of scientific theorizing? Which way will science—the physical as well as the human sciences—go?

1 Causality

The power of science lies in its explanations. Scientific explanation is based on causality. Since the rise of modern science the paradigm of a causal explanation has been provided by the laws of physics exemplified by Newtonian mechanics. The motion of a mass point described by a differential equation of the second order is the most precise formulation of the principle of causality. The various metaphysical interpretations of causality given by the classical philosophers have gradually ceded to the interpretations—equally metaphysical—given by the physicists. Since the revolution in epistemology (the theory of knowledge) started by Hume under the influence of Newtonianism, the principle of causality has remained in the centre of even purely scientific discussion.

The Newtonian method of mathematization was immensely successful and attained its final triumph with the completion of classical physics at the end of the last century. The logic of causality was seen in terms of the mathematical structure of the equation of motion. Given initial and boundary conditions, the position of a mass point at a later time t_2 is causally related to the position at the earlier time t_1. Thus, the movement of a planet is completely calculable and can be predicted without fail. In spite of Hume's criticism of the idea of necessity involved in causality, the formal, mathematical relationship between earlier and later position—that is, cause and effect—was incorporated into the semantics of the causal relation as determinism.

Necessity thus remained part of the pattern of scientific explanation within classical physics. This fact occasioned the violent disputes concerning determinism brought about by the statistical interpretation of the Schrödinger equation of quantum mechanics. Actually, determinism had already been curtailed to some extent by the special theory of relativity; but few people had noticed it. The finite speed of signal transmission restricts the reach of causal action. Accordingly, there exist at any given moment events that are causally neutral to each other. Determinism—the idea that any two successive positions of moving particles are necessarily related to one another—holds only for a selective and finite domain in relativity theory.

In quantum mechanics, as everyone knows, the bonds of determinism—in this strict, technical sense—are finally broken. Causality, however, in spite of the many confused views that have been voiced for almost fifty years, has remained. Schrödinger's equation, although expressing a statistical correlation, is still a differential equation of the second order. Given initial and boundary conditions, we can predict the future state of an electron in an atom from its earlier state, though only with a certain probability, not with necessity.

This change does not change anything in the logic of the causal explanation. There is general consensus that we explain a phenomenon if we can derive from the universal law—the differential equation—a particular statement, by inserting specific data concerning initial and boundary conditions, which correctly, that is numerically, describes this phenomenon. Testing and confirming the particular hypothesis, that is, the comparison of numerical values calculated and measured, validates the explanation.

This deductive scheme of explanation is, then, in the last resort based on Newton's equation of motion. Whatever the changes in the interpretation of causality, the logic has remained the same. However, the changes that physicists have been obliged to introduce were forced upon them by the demands of reality. Newtonian mechanics, and classical physics in general, give highly idealized descriptions of actual processes that, on closer inspection, cannot be maintained except, at best, as first approximations. This is illustrated by the well-known example of the process of measurement.

Quantum mechanics, with the principle of Uncertainty, has demonstrated that any measurement—involving the interaction of phenomenon and instrument—incurs an error which is never less than the quantum of action h. Thus, the initial data are, logically speaking, not determinate. The universal law is logically represented as an implication, for example, $(m)(f(m) \rightarrow a(m))$. In words: for all mass points, m, if a force acts on a mass point, then it is accelerated. If the premise of a universal implication is not strictly accurate, we must say that, logically, it is false, and from a false premise any consequent may follow, even a false one. Thus, the logical scheme of the causal explanation in terms of deduction has really become empty with quantum mechanics. The scheme only works within the context of idealization accepted in classical physics. Its explanatory power is, strictly speaking, nil. Do we not need to find another scheme?

Then two main arguments can be offered. Modern physics itself has shown, through the changing structure of its theories, how a new pattern of explanation could be obtained. Modern science in general, I mean the more recent human and social sciences such as psychology, also lead to and, indeed, demand another type of explanation than the one originating from Newtonian mechanics.

Of course, most scientists today—even the most mechanistically-minded among them—will agree that the concepts of physics cannot be imposed upon, say, psychology. Concepts that are appropriate for describing 'dead' atoms do not fit living human beings. All the same, there is a general conviction among most scientists that the causal type of explanation is the only one that can be envisaged. Is this really so?

Let us turn to physics again and take a quick look to see how relativity theory, quantum mechanics and symmetry physics suggest another explanatory scheme.

The first point to make is that physics has progressively shed the various idealizations with which it originally started. The artificial laboratory experiment allowed the physicist to simplify the 'natural' phenomenon and to abstract troublesome features from it. Thus the concepts needed to describe the laboratory phenomenon became easier to use, allowed strict though simple mathematics to be applied, and the mechanistic model of nature was so established. The model had the great virtue of providing objectivity, in the sense that the experimenter's wishes and fears, not even his experimental technique, could distort the result. Since measurement is an idealized process in classical physics, no reference to the scientist performing it is needed nor even allowed for. The human being has been exorcised in order to obtain objectivity.

This objectivity, however, was really quite precarious and could only be maintained with difficulty. Laplace's Superman was required to keep the mechanistic universe objective and Maxwell's demon had the same purpose of maintaining determinism within statistical mechanics. A very high price was paid for this spurious objectivity. Error was considered to be accidental, due to imperfections in the instrument or to the mistakes of the observer, in principle negligible and so outside the scheme of things. The necessary interaction between measuring instrument and phenomenon was disregarded. Thus, the actual uncertainty of the results could not even be computed. The illusion of the certainty and objectivity of our knowledge was created by ignorance.

The fear that the results of experiment would be distorted by acknowledging the participation of the scientist in the measuring process led to his total banishment from the scene of action. This extreme idealization, however, was gradually given up under the impact of new theories. Relativity theory re-introduced the human observer through the requirement of invariance. Quantum mechanics specified exactly the lower limit to the margin of error necessarily incurred in the measuring process. In neither instance do the personal fears and wishes of the experimenter enter into the result as was falsely alleged originally. In fact, our results are now more accurately known though less certain. Physics has shed idealization and become more realistic. We know more rather than less through acknowledging human participation. Even physics, like all sciences, is made by men.

This epistemological breakthrough must be recognized as being due to the loosening of the causal pattern of explanation. Any meta-scientific understanding must come from the actual theories and practices of science, if it is to be of any value. And, indeed, we can see how a new kind of relatedness gradually has come into physics. Relativity theory replaced the mechanistic universe by the universe of light. Light signals relate events to each other over finite distances and through finite times. Quantum mechanics provides statistical information through the wave function. A light signal, too, is the carrier of information since it is characterized not only by energy but also by entropy. It is capable of modulation and so the signal can carry a message. Finally, symmetry physics—as the present stage of the development of physical theory—forces us to look more closely at the process of deliberate experimentation which, after all, is the basis of modern science. The scattering of a beam of particles by another such beam is the only type of symmetry experiment; a source of high energy particles has to be carefully prepared. In practice this has always been needed though not acknowledged in the theory. A physical source is prepared with such properties as to be likely to give us the results we anticipate for theoretical reasons. In other words, we prepare a source of information and try to extract a message from it.

2 Information

The whole development of physics can be seen as a change from the concept of causality to the conception of communication and information. The causal law and the deductive scheme of explanation based on it has become more and more vacuous with successive theories. Today, the law is said to be, at best, 'an inference licence'. This licence applies only to the highly idealized model of a natural process. When regarded from the logical viewpoint, the law is a 'contrary-to-fact conditional' since the premise of the implication—the initial datum—is never factually true. The whole burden of explanation is put upon the application of this ideal to the real world and this depends on the skill and understanding of the scientist, very personal characteristics which this method was supposed to eliminate. The explanatory power of the causal explanation no longer prevails in modern physics. Causality, successful in Newtonian mechanics and useful within classical physics, has become a myth today. This fate usually befalls philosophical or meta-scientific, concepts. Scientific theorizing begins with metaphysics and, gradually, we shed the 'surplus meaning' of our concepts through the progressive cycles of abstraction and generalization until the concepts reveal their logical 'skeleton'. This has happened with specific concepts like that of atom as well as with general concepts such as induction and causality. The limitations of traditional, metaphysical induction have been widely recognized and its method replaced by mathematical statistics. The fact that we have reached the end of the usefulness of the metaphysical concept of causality is not generally accepted.

Yet both developments are parallel as is to be expected from their origins in Greek philosophy.

The emergence of the concept of information in the explanatory scheme of science involves a complete reorientation in our attitude towards nature. This, I think, is the main reason why so many people feel reluctant, if not unable, to give up causality. Once more, we must emphasize here that information is the natural successor to causality, the 'next higher approximation' in the series of conceptual changes within science. The *correspondence* principle that governs the succession of scientific concepts does also work here within the context of meta-science. While causality considers the relation between events in terms of energy alone, communication (or information) adds to it *entropy*, or *order*. Thus, information theory introduces a new dimension, over and above the four space-time co-ordinates, for describing a natural process. The theory allows us, therefore, to refer to the human participant—the experimenter—in the acquisition of data and even to refer, though in a very restricted sense, to the meaning of the message which is communicated. It is exactly this more comprehensive conception that is needed to restore power to our explanatory schemes which the causal concept has lost in modern science.

How can this change in attitude be described quickly? First of all, we have to speak in more detail about what the scientist does in the laboratory. The scientist designs and performs an experiment in order to extract new information about a particular state of affairs. Thus he has to prepare a source of information—some special apparatus like a high-voltage machine. This source, according to the theoretical understanding of the scientist, is to represent a closed system capable of many, and as yet unknown, states. It is as highly organized a system as possible, viewed macroscopically and externally, whilst it contains a great deal of uncertainty, that is, potential information, among its microscopic and internal components. The experiment consists in stimulating an information flow passing from the source, through a suitable channel, to a receiver. The artificial state of affairs created by the scientist in the laboratory represents, then, past knowledge and new (potential) information. The scientist's knowledge and imagination decide what kind of information source he constructs and what sort of information he extracts from it.

This human participation in the process of experimentation does put an unavoidable limitation on it. The scientist's theoretical knowledge and practical skill will affect the outcome of the experiment and may even distort or falsify the result. We cannot assess this personal factor numerically but we must not neglect it in our account of scientific work, for it is here that the creativity of the scientist is exercised. More than that, we have to recognize that, although tested and confirmed by experiment, all our theories still carry the imprint of the various scientists who invented them. If Descartes and Leibnitz had won out instead of Newton, would we have the same kind of mechanics today? No doubt, basic laws—like the three of Newton—would also exist in a Cartesian mechanics but their formulation would be very different and physics would have developed differently, too. We believe unthinkingly that science is the impersonal product of a person, the scientist. We are brainwashed into believing that, whoever the experimenter, the same concepts would be invented and the same results obtained. This is a comforting view—as if human stupidity and malice that affect every endeavour could not touch scientific work. It is to forget the natural origin of human inquisitiveness, including the search for knowledge, in the unconscious and infantile strivings which, after all, show also in magic, witchcraft and other superstitions. This idealization of an ability that makes the scientist into a perfect performer has to be given up just as the corresponding idealized view of measurement without error had to be abandoned.

Now, let us turn to the more technical aspect of information theory. The knowledge incorporated into the information source is measured by the redundancy of the message transmitted from it. The amount of information—the increase in knowledge—is measured by the 'surprise value' it has to the experimenter. This amount is mathematically formulated as $\sum_i p_i \log p_i$—the statistical expression that indicates an unexpected order in the sequence of symbols occurring in the message. The order is unexpected relative to the standard—the known order of symbols within the source—set by the experimenter. Although the source and the message transmitted from it need not possess the same order (or disorder), the message can certainly not be more ordered than its source. Of course, we are interested in the disorder, or uncertainty, in the symbol sequence for it represents (potential) information.

The extraction of information from a source is described in physics by the theory of measurement. Noise is the inevitable accompaniment of the process. Once more, human intervention is crucial: for the experimenter combats the noise by suitable 'coding' of the signals from the source to produce a message that can reach the receiver with minimum distortion. Of course, it means that the noise-entropy must be compensated somewhere else, outside the actual experimental system, that is, in the power supply or in the surroundings generally. Not only energy but order as well is required to obtain information. In the last resort, it is the human being who, by applying his knowledge re-introduces order and restores the balance. If he is not able to do so, the experiment cannot be repeated and no reliable datum could result from it. Without repetition there is no knowledge.

Thus, human participation is needed in every step. Gradually, a completely different idea emerges of what constitutes scientific research. Bacon already had said that we 'put nature to the question', that is, interrogate her in an artificial way. Now we must enlarge this conception and say that we produce artificial situations in which the external phenomenon and the internal, mental-emotional, constitution of man share equally. This interaction replaces the one-sided action previously assumed as the effect of the external phenomenon. In consequence, causality has to be replaced by communication (or information flow). Thus, the human being is back in the centre of the action where he belongs.

We can then no longer describe the acquisition of knowledge as the investigation of a ready-made, totally self-contained, external world that is probed from the outside by the scientist. We can no longer even say that nature is an entity totally independent of us and that it behaves according to immutable laws. All laws are man-made in the sense that they express, and are established, only through human intervention. Here we might remind ourselves that the original conception of causal law, in Greek philosophy, was abstracted from the moral law of revenge exercised by the Erinyes. The experience of personal motivation lies underneath the impersonal and 'objective' principle of causality. The concept of law—outside science—has never lost its man-made character even though society as a whole rather than the individual has put its imprint on the law. Within science, however, the law has been made into the very centre of our impersonal and 'objective' research, the expression of a nature that exists externally to, and independently of, ourselves. And yet, even physics is a 'human' science in the sense that it represents knowledge produced by human beings and inevitably reflecting, therefore, human abilities of perception and cognition. Brain organization, for example, and the structure of the eye that have arisen through the long periods of human evolution have fitted us to interact with our physical environment, and with other human beings, so as to acquire some knowledge about them. This knowledge is not handed down to us, in pristine and absolute form, like the Ten Commandments. It is acquired by hard work that, at every stage, incorporates human error and misconception. And only by slow and strenuous effort can we free science from these defects and gradually improve its accuracy and enlarge its scope. This is the essence of scientific research that cannot be described or evaluated without reference to the human observer or experimenter.

Scientific explanation is therefore concerned with human activity, not with the workings of an inhuman, physical nature, even when we do physics. Physics is what physicists do in their laboratories—not the description of an ideal, external world, of a perfect machine that ticks over by itself without our help or interference. People say, 'It is a law of nature', and they imply that nothing can be done about it, that whatever is expressed by it represents really 'super-natural', all-powerful authority. This is the metaphysical, if not to say theological, attitude that lies beneath the conception of causal law. As my argument has shown, I think, causality as a principle of scientific explanation has become exhausted. There is a metaphysical residue within it that makes application in modern physics, and on a more abstract level, impossible. And there is its logical skeleton as universal implication which, at best, may be interpreted as 'inference licence', at worst, as contrary-to-fact-conditional. We still, of course, explain the unknown by the known—the past by the present; and we predict the future on the basis of past and present knowledge. But the explanation is concerned with experimentation—with the scientist's work in the laboratory. The universal law is then no more than a rule that the scientist follows when carrying out an experiment, and its universality is not ontological, but simply expresses the fact that we may use the rule indefinitely. A hypothesis is then concerned not with discovering a hidden feature of the mechanism but with establishing a new rule for experimentation. The rule tells us how to experiment in order to obtain new information.

For example, in symmetry physics the so-called parity experiment is designed to test the possible existence of an 'anti-world'. The electrons in the β-decay of radioactive cobalt are found not to change their direction when other characteristics in the experiment, for example, the direction of the external magnetic field, are changed. Viewed through a mirror, that is, as a phenomenon in the 'anti-world', the spatial characteristics of the electron beam do not change with reflection as we expect. Parity—which is the property of any object or process such that its mirror image is equally allowed—is broken in this instance. This lack of symmetry (for a specific process, for example, weak nuclear interaction) between the actual world and the anti-world of elementary particles is then found through following a number of rules in experimentation. The experiment is so designed as to be able to exhibit a certain symmetry among the properties of elementary particles. (If we want to keep the symmetry in the anti-world, we have to augment the operation of co-ordinate inversion by an operation called 'charge conjugation' in which the negative electrons are changed into positrons. Then, the 'combined' operation is 'invariant' under inversion. Co-ordinate inversion without particle conjugation is a valid symmetry operation only if we exclude weak interactions from our considerations.)

Human activity rather than an external nature is the subject-matter of physics; and rules of experimentation rather than universal laws describe it. It is necessary to widen our explanatory schemes and to consider the rôle of the experimenter even in physics. This will be even more necessary in the social sciences where the live human being rather than the dead atom is under investigation.

3 Motivation

To some extent this rôle of experimenter has been

recognized when motivational rather than causal explanations are offered in the study of human behaviour. However, this recognition is often only nominal. For what do we mean by 'motive'? There is the general definition of 'motive' as a disposition of the organism to act in a certain way. But how do we describe this disposition? It is not a simple physical thing that is exhaustively described by its four space-time co-ordinates. For this is what people, often inadvertently, do when they try to account for a motive by referring to the overt behaviour of the physical organism or, perhaps, to its brain processes. 'Motive' is, however, not another word for cause; nor is a disposition a kind of internal cause. A living organism represents a very much higher level of organization than even the most complex, physical, dead system. We need not here appeal to mysterious, and logically very objectionable, occult properties which, for instance, are postulated by vitalism. All we have to say is that a highly organized system possesses a high content of (potential) information which it has been able to acquire by virtue of its structure. Human beings, after all, are born with some innate abilities and they improve these and develop new abilities in the course of their development. A motive is, therefore, to be identified with the instruction a human being may possess and according to which he may, or may not, act. Instruction is (potential) information acquired as the result of past experience. This is very unlike the cause, that is, a definite set of initial and boundary conditions, from which a movement must follow by necessity.

Movement, of course, is also a concept that is inadequate for describing human behaviour. For the physical, overt behaviour represents only a part of what a human being does. A motive, conscious or unconscious, is acted on by him only in specific conditions, for a purpose. A human being, unlike the atom, is a purposive organism. He is endowed with self-movement and can act according to a purpose. His behaviour is to be described as action rather than mere movement. And he justifies his action by giving reasons for it rather than merely explaining it by stating the physical antecedent conditions for the physical movement, if any, which is involved in the action. There is an extra dimension to be considered in human action, over and above the four space-time co-ordinates, through which action can be distinguished from movement.

This extra dimension arises from the greater degree of freedom that a highly organized system must be said to possess in comparison with a less organized system. It is, once more, not a mysterious or supernatural characteristic : the fifth dimension simply represents the ability of the human being to act in order to achieve, consciously or unconsciously, a certain purpose. Being hungry, we go to the kitchen and find something to eat. The physical movement alone does not describe what we do. And we cannot hope to achieve a satisfactory explanation of human behaviour unless we describe all the parameters that enter into it. The push-pull model of mechanics is not adequate since it does not encompass purpose.

Purpose, unfortunately, has often been considered a dirty word as if it were, by its very nature, unscientific. It is true that purposes may be misinterpreted and that we can lie about them. We can, however, be equally mistaken or dishonest about causes. Nor is it correct to consider purpose teleologically. That is to say, we regard a purpose as a cause in the future and retrodict present movement from it. Naturally, we come to grief if it turns out that the purpose is not achieved in the end. Of course, it is the idea of a desirable future state of affairs rather than this actual future that stimulates our action. Purpose is anticipation; and we are back within the domain of motivation when interpreted as potential information or instruction.

Purpose brings in the idea of rule, and even of moral rule. For a purpose depends, within limits it is true, on our choice and decision: and we follow certain rules to achieve a definite purpose. Morality here is simply the natural ability of human beings to choose between alternatives; it is not the imposition of a specific code of behaviour which is conditioned by social and other requirements. Human beings have, by nature, a moral sense just as they have a sense of sight or smell. The sense of sight or smell does not determine a specific preference for a particular colour or odour though there is a range within which the human organism is capable of response. The same holds for the moral sense. Men are agents but there is a limited range of actions that human beings can carry out.

Motive and purpose, however, are not isolated, causally functioning items. They are not 'in the head': they have to be ascribed to the organism as a whole.

We always talk, therefore, about a system. In other words, we speak about a human being who is in a particular situation or state; and the situation he finds himself in is the result of his own growth and development and of social interaction. There is a history—during which his mental and emotional life has developed and, with it, the motives and aims, the reasons and purposes, that lead him to act. These are natural tendencies, accompanying biological growth and mental-emotional development, that affect everybody's actions. Evolution, whether biological, mental-emotional or social—though working in a very different manner in each instance—suggests the model of man in terms of motive, purpose and rule-following behaviour.

For we try to fulfil a purpose or work out a motive, by following a certain course of action—consciously or unconsciously; that is, we follow a rule. If we identify motive with instruction, then it is natural enough that we follow this instruction. The instruction is the result of past experience—we may misunderstand it or, indeed, it may arise from misinterpretation of our experiences. We still follow a rule which is inherent in the instruction. It is not a law of nature, a causal process since choice and decision are involved, in principle. We are not forced to act in the way we do act except in extreme circumstances. We can, at least sometimes, inhibit our motivation and control our behaviour.

Purposive behaviour is not outside the scope of scientific explanation; and it is the essential characteristic of human personality without which no science of psychology can be constructed. Purpose, however, is a scientific concept when it is interpreted in terms of potential information (like motive), but not in terms of an inverted causality, as traditional teleology suggests. For we need to describe human behaviour from within, not only from without: we are agents and not only onlookers. Thus, we must be able to include inner experience in the scientific description of human personality and accept it as a legitimate source of information. It is usual to speak here, for the sake of clearer distinction, of understanding ourselves and other human beings, in contrast to knowing human behaviour from the outside, through observation. For it is quite legitimate to make use of our personal experience, in order to understand other people. Whatever the differences between us, we are—some better than others—at least in principle quite able to put ourselves into another's place and feel what it is like. This ability to identify, to project and introject, is a fundamental human trait and there is no reason to deny it in our account of how we acquire information about ourselves and other people. *Anthropomorphism*, in a limited sense, is quite justified in psychology, though not in physics; and, equally, *physikomorphism*—although falsely taken as a scientific standard by some—is quite out of place in psychology.

This inner experience and the understanding that comes from it, is, however, quite different from the introspection of the philosophers. We do not look into ourselves as if we were outside observers, in order to objectify our inner world. As participants in the process we experience, consciously or unconsciously, relationships with other human beings. All human experience, in the first place, arises from interaction with other human beings; and each experience is incorporated, often in a distorted form, and transferred or applied to other experiences. (It may be 'depersonalized', as the example of causality shows.) This is the origin of human motivation: the information that we have stored up within us. Information, by definition, includes the human being as participant. Inner experience and understanding are primary processes. Experience of the external world and knowledge are, genetically, secondary since they can arise only after the human infant has had sufficient inner experience to project and construct such a world. The very general and relatively unspecific structure of inner experiences, the very plastic tendencies for the satisfaction of needs with which the human infant starts his life cannot, perhaps, be said to constitute a complete inner world. However, the many and rapid interactions of the baby with his mother quickly build up such an inner world. After a while the infant learns to separate out from this result of interaction what comes from the 'other' and what is 'himself'. Experience of the inner world and knowledge of the external world thus develop together. But since the inner world of an infant incorporates, if only as phantasy, the external world or, rather, the mother, who is its first representative, we are enabled to put ourselves in someone else's place, to experience what it is like to be another person, to know from the inside another's feelings, to understand them. This is a basic ability of man as manifested in our mental-emotional growth.

Once we recognize that understanding, like knowing, is a natural ability of ours that may be fostered or stifled, then a scientific description of human behaviour may, nay must, include it. This acceptance of the rôle the inner world plays in the process of knowledge implies, too, that *morality is implicit in any scientific explanation.* It is not true that any personal likes or dislikes are necessarily included in science; on the contrary, all our efforts, as scientists, are directed to avoid 'personal' involvement and distortion. Morality comes into science, in principle, when we regard science as a human activity for discovering reality and truth; and these are values since they give direction to our research. Science has a purpose and, in this general sense, it involves morality. Moreover, all morality requires us to free ourselves from—or at least be aware of—distorting phantasies; for no action can be moral unless the situation is seen correctly, as it exists in reality. Infantile phantasies result in infantile morality. Thus the best moral judgement available—the 'mature' morality of a healthy adult or the best approximation to it we can reach—is implicit in, and needed for, science. For science provides us, at any given time, with the reality of the world we live in and guarantees the truth of the statements about it.

Science as the activity of experimentation (and communication with other investigators) is thus described by rules rather than by laws. The model of scientific activity is not the 'impersonal' exploration of a passive nature but the human activity of creating new situations and relationships. All explanation has the aim of relating the novel experience to past and known experiences, of transforming the unfamiliar to the familiar. The familiar situation is expressed by the model (or full-blown theory, if we have one) into which we try to fit the unfamiliar phenomenon. The old mechanistic model of planetary motion and universal laws is no longer workable; instead, we need a model of human action and general rules for explaining science.

The explanation of human action is necessarily different from the explanation of physical movement. Different modes are involved and different purposes or aims; and the situations to be explained differ completely in their nature. The physical situation is artificially produced in the laboratory and is the result of human action, of experimentation. Thus, the explanation of action is seen at once to be more comprehensive than the explanation of physical movement. *Causal explanation is a special and narrower type which is included in the rational explanation of human behaviour.*

The explanation of human behaviour, moreover, must refer to natural, not artificial laboratory, conditions. Human action cannot be simplified, idealized, abstracted and universalized as physical movement can. What we want to explain, after all, is our behaviour here and now, not a simulation of it. Animal experiments can shed light on human behaviour, of course; but they cannot provide an adequate explanation for it. The same holds for any kind of experiment under controlled conditions: they cannot replace the total situation in which a human being lives and acts.

Finally, of course, the purpose of an explanation of human action is quite different from the purpose of the causal explanation of physical movement. Causal explanation is inextricably bound to prediction; and prediction refers to space-time movement. But we want to explain, in psychology for instance, a character or a present behaviour pattern. We are not interested, nor does it make much sense, to predict the future movement of the physical body of a person. While the physical body may, or may not, react to a stimulus, the person reacts to the meaning of a situation, the meaning it has for him. Thus, we must relate a person's present behaviour, and his interpretation of the situation, to his past experiences within the context of a general theory, or model, of human development. In order to bring out the difference between causal and motivational, or rational, explanation we usually say that we want to understand rather than to explain, someone's behaviour.

We understand an action or situation because we can

put ourselves in the place of some other human being. This is the great advantage the human sciences have over the physical sciences. Without being able to identify, without being able 'to feel in our bones', what another person's experience is like, we could not make progress in the human sciences where we cannot impose artificial laboratory conditions. Past experience, too, builds up the understanding of ourselves since 'motor memory' allows us to recall situations similar to those we want to understand. Non-verbal and direct—though still corrigible since, in principle, it can be mistaken—our understanding is in fact prior to the verbal and discursive knowledge which we accept as impersonal and objective. This must be so genetically; an infant understands rather than knows until he is old enough to reason. This must be so generally for, in the last resort, we must be moved to some activity, our feelings and emotions must be engaged, if we have really incorporated whatever there is to know. Understanding is a necessary part of all scientific activity. It represents the human factor which can never be eliminated nor even temporarily suppressed anywhere. Physics is an example which illustrates this.

The rejection of understanding as a source of information has been based on the atomistic theory of perception and introspection. If introspection means the outside, spectator observation of inner processes then, of course, there is no such thing as understanding. Inner experience, however, exists—and is even in some sense prior to external experience. This is the inside, participant's feeling. We rely on it as genuine if there is integration—whatever we have understood makes the situation more coherent and consistent. Indeed, the criterion of integration, when 'things fall into place', the 'click of recognition' is the hallmark of understanding. All science is based on experience but not necessarily on observation and perception. It is the outsider view of associationist psychology—of atomism applied to human personality—that has made people reject understanding as source of information. It is because we have felt what a certain situation is like that we can interpret what we observe in some other person. We should have no basis for outsider observation if we had never experienced the insider's feeling due to being a participant in the situation, or at least a similar situation, in question. We can always imagine ourselves in a novel situation. But the conviction which this carries with it is very much less than the conviction of memory. Psychology and all social or human sciences, have as their subject-matter the human being. We can gain no knowledge of another person if we regard it as a problem of perception, if we restrict ourselves to the outsider's view. First-person understanding is the prerequisite of third-person knowledge. The artificial and arbitrary constructions of physics have provided us with the wrong theoretical model and cultural paradigm of the process by which knowledge is acquired.

4 The future of science

Science, and the technology based upon it, is the most powerful force in modern society. Science is the result of the natural, human drive for knowledge, of our *epistemophilia*. Like all natural phenomena, science changes and develops. Animism—the most primitive attempt of man to orient himself in the world—was superseded by atomism which represents the next stage in this development, that is, the discovery of an indestructible matter. Atomism led, in the Renaissance, to the mechanistic interpretation of nature. The mechanization of the world picture was completed with Newton and brought with it the possibility of constructing simple machines—the pump and the steam engine—as sources of power; this represents the third stage in the development of science in which mechanical power and energy are discovered. It brought about the first Industrial Revolution. Its socially objectionable consequences—exploitation of labour and alienation—were, to a large extent, caused by the feudal and commercial organization of society then existing. To a lesser extent, the mechanistic character of science and technology was conducive to the neglect of the human factor. Indeed, the objectivity of science was incorrectly attributed to the elimination of the human experimenter from the scene of action. When this false absolutism was abolished by Einstein and when the quantum physicists made the experimenter the necessary condition for acquiring knowledge, a fourth stage in the development of science was reached. Information (or negative-entropy), already implicit as a concept in nineteenth century physics, rather than energy (and causality) has become the basic characteristic of a natural process. This has brought about the second Industrial Revolution which began at the end of the last war and is still continuing. Information brings back the human factor by necessity. Information flow, or communication, improves the cohesion of mankind as a single community. If one regards the growth of science as part of human evolution, one can at least hope that humanization—rather than mechanization— is the next stage in the development.

Summary

Methodology is the self-criticism of the scientist. Thus it reflects the basic attitude, the inner motivations that produce contemporary science. All explanation consists in relating the unknown to the known, making the novel situation familiar, and so extending the pattern of relationships between phenomena, producing order out of chaos. It depends therefore on the conceptual framework within which this pattern is to be established, on the theory (or theoretical model), and on the cultural paradigm accepted at the time. The causal model of natural phenomena is accepted even today as the basic pattern. But this model is derived from mechanics, that is, Newton's equation of motion, which provides the most refined interpretation of cause as initial and boundary conditions. Atomism is an essential part of the model, that is, the existence of separate, isolated objects. And prediction of movement for a later space-time point, if successful, confirms the explanation. Semantically, the mathematical equation becomes the universal law from which a particular hypothesis concerning the future can be logically derived. This is the H-D scheme of explanation, the only scheme generally accepted today.

Unfortunately, the scheme has many weaknesses and it no longer suffices to explain modern physics. Logically, the universal law is a contrary-to-fact conditional and so unverifiable. At best, it may be taken as rule of experimentation, as an 'inference licence'. Moreover, relativity theory and quantum mechanics have rejected the absolute and necessitarian character of the universal law. The human factor—suppressed, though never really eliminated in the model of classical physics—had to be re-introduced openly. Science, at long last, is recognized as an activity—experimentation represents the beginning of the modern era, after all—and it is no longer a book of words. The scientist plays an essential rôle in the acquisition of data and not only in the formulation of a theory or hypothesis. Communication between scientists, the chain of information—the exchange of a signal between events represents the general pattern of the real phenomena.

This brings us nearer to explanation in the 'human' sciences. Here we have action rather than movement; reason, purpose or motive rather than cause; and 'life forms' rather than future space-time positions have to be explained. Men are agents, participants: they can move by themselves—in contrast to atoms that move only under external force. Meaning, rather than physical stimulus, is the unit of action. 'Rule-following' rather than 'law-governed' is the concept which describes human behaviour. Rule-following—language is the prime example—is learned and corrigible, not the effect of a conditioned reflex. Rule-following remains under the agent's control, at least more or less; thus it is responsible behaviour or moral. For learning pre-supposes interpersonal communication, human relationship. Man is a social, and *a fortiori* ethical, animal. Causal movement belongs only to the infra-structure of rational action.

Every situation requires an evaluation, that is, the appraisal of its value for the agent, and also a method of approach, that is, the goal of an action. Thus the model or paradigm of human behaviour is very different from that of Newtonian mechanics. Our explanations are with respect to this model of human behaviour and are seen to be valid if integration rather than prediction is taken as the criterion. Science has to assume a non-Euclidean type of theoretical organization in which the logical, or deductive, hierarchy of statements is replaced by the hierarchy of order and information content of meaningful message activity. Information, message and meaning are concepts that necessarily presuppose man as participant in the scientific process. Once more, as in classical Greece and in the Renaissance, science is the expression of Humanism.

About the authors

Dr D. J. Stewart first graduated in philosophy and psychology, and was for some years a university lecturer in psychology, but is now on the staff of the post-graduate Department of Cybernetics at Brunel University, lecturing on cybernetics and philosophy and supervising research projects in areas as diverse as business management, poetry, police work, psychiatry and town planning. Rationalism is a long-standing interest of his; he has been a director of the Rationalist Press Association since 1963 and its Chairman for eight years. He is an Associate Editor of the *Journal of Moral Education* and a member of the Editorial Board of *Question.*

Dr Christopher Evans is an experimental psychologist who has worked in human perception and Man-Computer Interaction. He has published over 50 scientific papers, and is the co-author of a theory which compares dreaming to computer programme clearance. He is Secretary, and one of the founder members, of the Brain Research Association. Dr Evans believes passionately in the importance of effective communication between scientists and non-scientists and devotes much time to furthering these causes. He lectures extensively throughout Britain and appears frequently on television and radio. His book *Cults of Unreason* is to be published by Harrap shortly, and he is currently writing one on dreams—*Landscapes of the Night.*

Colin Campbell is a lecturer in Sociology at the University of York. He is the author of *Towards a Sociology of Irreligion* and Vice Chairman of the Rationalist Press Association.

Professor E. H. Hutten is Professor of Theoretical Physics at Royal Holloway College, University of London. His books include, among others, *The Origins of Science, The Ideas of Physics.* Professor Hutten is on the Editorial Advisory Board of *Question.*

DATE DUE
